MATHEMATICS WITH BUSINESS APPLICATIONS

MATHEMATICS WITH BUSINESS APPLICATIONS

Edited by:

Ivan Stanimirović

ARCLER PRESS

www.arclerpress.com

Mathematics with Business Applications

Ivan Stanimirović

Arcler Press
2010 Winston Park Drive,
2nd Floor
Oakville, ON L6H 5R7
Canada
www.arclerpress.com
Tel: 001-289-291-7705
001-905-616-2116
Fax: 001-289-291-7601
Email: orders@arclereducation.com

ISBN: 978-1-77361-071-9 (Hardcover)

DECLARATION

Some content or chapters in this book are open access copyright free published research work, which is published under Creative Commons License and are indicated with the citation. We are thankful to the publishers and authors of the content and chapters as without them this book wouldn't have been possible.

ABOUT THE EDITOR

Ivan Stanimirovic gained his PhD from University of Niš, Serbia in 2013. His work spans from multi-objective optimization methods to applications of generalized matrix inverses in areas such as image processing and computer graphics and visualisations. He is currently working as an Assistant professor at Faculty of Sciences and Mathematics at University of Niš on computing generalized matrix inverses and its applications.

TABLE OF CONTENTS

List of Contributors xiii

Preface .... *xvii*

Chapter 1 Mathematical Modeling of Economic Growth 1

1. Importance of Modeling of Financial Information 2

Chapter 2 Mathematical Modeling of Loan Repayment 13

1. Amortization System Constant (ASC) 14
2. French System of Price Amortization - (FSA) 15
3. Comparative Analysis of ASC and FSA Methods 15
4. American System of Amortization (ASA) 16
5. Application In Net Present Value 17

Chapter 3 Mathematical Models in E-Business 19

Chapter 4 Economic Concepts and Their Application in E-Business 25

1. Microeconomic Theory of The Firm 25
2. Cost Coordination 30
3. Transaction Costs 40
4. Agency Costs 43
5. Change Cost 48
6. Network Externality 51

Chapter 5 Mathematical Design of E-Business 55

1. Designing Organizational Structure 55
2. Product Design and Pricing Policy 63
3. Using Mathematics to Generate Captive Customers 69
4. Development of Client Networks 73

Chapter 6 Mathematical Models In Capital Management 79

1. Cash Management 82

2. Operation In Practice of Effective Management of Cash 84
3. Cash Conversion Cycle 85
4. Models of Demand For Cash Transactions. 86
5. Baumol Model 86
6. Cash Management Model Miller-Orr 87
7. Commercial Discounts 89
8. Inventory Control 89
9. Model EOQ 90
10. Discounts For Wholesale 91
11. Inventory Control With Uncertainty 92
12. ABC Inventory Control System 93
13. MRP Systems Subject (PNM) 93
14. Inventory Systems Just In Time (JIT) 94
15. Accounts Receivable Management 95
16. The Decision Granting of Basic Credit 97
17. Sources of Credits 98
18. The Five C Credit 98
19. Accumulation Accounts Receivable 99
20. Short-Term Financing. 100
21. Cost of Capital And Its Implications From Business Investment. 105
22. Capital Cost 106
23. Costs of Sources of Capital 106
24. Cost of Capital 107
25. Average Cost of Capital 107

Chapter 7 Personality Traits as Predictors of Shopping Motivations And Behaviors: A Canonical Correlation Analysis 111

Abstract 111
1. Introduction 112
2. Method 120
3. Results 121
4. Discussion 126
5. Managerial Implications 129
References 130

Chapter 8 **The Relationship Between Growth-Inequality-Poverty Triangle and Environmental Degradation: Unveiling The Reality....137**

Abstract....137
1. Introduction....138
2. Literature Review....143
3. Data Source and Methodological Framework....151
4. Results and Discussions....158
5. Conclusion and Policy Recommendations....163

References....167

Chapter 9 **The Importance Of Corporate Brand Identity In Business Management: An Application To The Uk Banking Sector....177**

Abstract....177
1. Introduction....178
2. Conceptual Framework and Hypotheses....183
3. Method....187
4. Analyses and Results....190
5. Conclusions, Implications and Directions For Further Research....194

Acknowledgements....197

References....197

Chapter 10 **The Evolution of Strategic Management Research: Recent Trends and Current Directions....203**

Abstract....203
1. Introduction: Trends In Strategic Management Research....204
2. The Evolution of Strategic Management Research: The Dual Pendulum....207
3. The Tension Between A Firm's Interior And Its Environment....209
4. The Tension Between Macro And Micro Levels....212
5. About The Monograph....215
6. Conclusion and Final Reflections....218

References....219

Chapter 11 **The Institution-Based View Of Strategy: How To Measure It....225**

Abstract....225
1. Introduction....226
2. The Institution-Based View of The Firm....228

3. Measuring Institutions 234
4. The Measurement Of Informal Institutions 250
5. Global Leadership and Organizational Behavior Effectiveness (Globe) 253
Conclusions 261
References 264

Chapter 12 Investigating The Relationship between Corporate Social Responsibility And Earnings Management: Evidence From Spain 273
Abstract 273
1. Introduction 274
2. Literature Review And Hypothesis Development 277
3. Research Design 280
4. Results 288
Conclusions 293
References 295

Chapter 13 Relationship Between Corporate Social Responsibility and Competitive Performance In Spanish Smes: Empirical Evidence From a Stakeholders' Perspective 305
Abstract 305
1. Introduction 306
2. CSR and Firm Performance Link 308
3. The Direct Relationship Between CSR And Firm Performance 309
4. CSR-Firm Performance: The Mediating Role of Relational Improvements 312
5. Methods 314
6. Measures 316
7. Discussion 332
Conclusion, Limitations, and Future Research 334
References 338

References 347

Citations 351

Index 353

LIST OF CONTRIBUTORS

Ivan Stanimirović

Ali Gohary
Department of Business Administration, Qazvin Branch, Islamic Azad University, Qazvin, Iran

Kambiz Heidarzadeh Hanzaee
Department of Business Management, Science and Research Branch, Islamic Azad University, Tehran, Iran.

Syeda Anam Hassan,
Department of Management Sciences, COMSATS Institute of Information Technology, Abbottabad, Pakistan

Khalid Zaman
Department of Management Sciences, COMSATS Institute of Information Technology, Abbottabad, Pakistan

Shah Gul
Department of Management Sciences, COMSATS Institute of Information Technology, Abbottabad, Pakistan

Isabel Buil
Departamento de Dirección de Marketing e Investigación de Mercados, Facultad de Economía y Empresa, Universidad de Zaragoza, Edificio Lorenzo Normante, C/ María de Luna s/n, 50018 Zaragoza, Spain

Sara Catalán
Departamento de Dirección de Marketing e Investigación de Mercados,

Facultad de Economía y Empresa, Universidad de
Zaragoza, Gran Vía 2, 50005 Zaragoza, Spain

Eva Martínez
Departamento de Dirección de Marketing e Investigación de Mercados, Facultad de Economía y Empresa, Universidad de
Zaragoza, Gran Vía 2, 50005 Zaragoza, Spain

Luis Ángel **Guerras-Martín**
Universidad Rey Juan Carlos, Spain

Anoop Madhok
Schulich School of Business, York University, Canada

Ángeles **Montoro-Sánchez**
Universidad Complutense de Madrid, Spain

Elisabet Garrido
Universidad de Zaragoza, Violante de Hungría, 23, Zaragoza, Spain

Jaime Gomez
Universidad de La Rioja, Cigüena˜ 60, Logrono, ˜ Spain

Juan P. Maicas
Universidad de Zaragoza, Gran Via 2, Zaragoza, Spain

Raquel Orcos
CEU San Pablo, Julián Romea, 23, Madrid, Spain

Ester Gras-Gil
Department of Accounting and Finance, Faculty of Economics and Business, University of Murcia, Spain

Mercedes Palacios Manzano
Department of Accounting and Finance, Faculty of Economics and Business, University of Murcia, Spain

Joaquín Hernández Fernández
Department of Accounting and Finance, Faculty of Economics and Business, University of Murcia, Spain

Jesús Herrera Madueno
Financial Economics and Accounting Department at University of Cádiz, Business and Administration College, Av. Duque de Nájera, 8, 11002 Cádiz, Spain

Manuel Larrán Jorge
Financial Economics and Accounting Department at University of Cádiz, Business and Administration College, Av. Duque de Nájera, 8, 11002 Cádiz, Spain

Isabel Martínez Conesa
Financial Economics and Accounting Department at University of Murcia; Business and Administration College, Facultad de Economía y Empresa, Campus de Espinardo, 30100 Murcia, Spain

Domingo Martínez-Martínez
Financial Economics and Accounting Department at University of Cádiz, Business and Administration College, Av. Duque de Nájera, 8, 11002 Cádiz, Spain

PREFACE

We define different varieties of business conducted over the Internet or supported by this technology, which are called e-Business. These businesses can range from sell or Internet auctions to manage internal business processes -distribution, production, supply, finance, etc. using such technology.

Differential equations are a powerful and versatile tool for solving problems from the most diverse horizons: from mechanics to biology, economics electricity, etc. The first step in solving these problems is modeling, ie the "translation" in mathematical relationships of the most important of the situation in question intrinsic aspects.

The mathematical treatment of a problem from technical or other branch of science can be a difficult job, it is the combination of two different domains: the phenomena being studied (financial, ecological, economic, etc) and mathematics.

Solving a problem basically comprises the following steps:

1) Determination of intrinsic qualitative aspects of the problem: choosing a reference system of independent variables and unknown variables.
2) Writing equations. These essentially they translate laws or physical, biological, economic behavior, etc.
3) Processing and study of the equations obtained. Analytical or approximate solutions of the unknowns are being looked for.
4) Interpretation of the mathematical results by comparison with experience.

We understand mathematical model to approximate representation of a phenomenon by mathematical relationships (usually equations, differential equations in our case). This approach is appropriate if the conclusions that can be drawn have sufficient similarity to the observed phenomena. Its formulation corresponds to steps

1) and 2) mentioned above. It is virtually impossible for a model to represent all facets of the phenomenon. Therefore we insist on the need to take into account the most relevant aspects of the situation in question.

Sometimes the same model can be reformulated facilitating its resolution. As well as the equations are also involved data models, inaccuracies in measuring these entail errors in the results.

In most cases there is no analytical solution of the mathematical model by which we must pass numerical modeling. This is another source of inaccuracies and in the context of numerical analysis can establish error bounds between the solution of the numerical model and the exact solution of the mathematical model. The mathematical model is disturbed numeric character as mathematical operations as the derivation or integration are not accurate.

The numerical model for its size or complexity may require computer processing involves the disturbance arithmetic, since the representation of real numbers and elementary arithmetic is about.

In this book, we will try to figure out how the essential bookkeeping condition is utilized to keep all bookkeeping records in adjust. We broke down business exchanges to perceive how they influenced each piece of the bookkeeping condition. Money, credit, income and cost, withdrawals, and speculation exchanges were inspected for a specific business. It should be kept in mind that a solid comprehension of the fundamental bookkeeping condition is the way to understanding the impacts of any business exchange. In conclusion, the resolution of a problem for numerical computation, the result cascading different types of errors, rarely produces an exact solution.

Editor

CHAPTER 1

MATHEMATICAL MODELING OF ECONOMIC GROWTH

Ivan Stanimirović

Mathematical models can be classified according to different criteria.

- If the data are affected by chance (which mathematically measured by the odds) model is stochastic. Otherwise, deterministic model is said.
- The model will be linear if the functions and conditions are linear, but speaks of nonlinear model.
- The continuity of data variables and make the model is continuous, and its opposite the discrete model.

As there is a more complete theory under the assumption of linearity and better adapted numerical methods, linear models are "easier" to solve. Abusive linearization nonlinearities can lead to disastrous results. Stochastic models can receive treatment to return to them deterministic.

In recent times, driven by information technology, biology and neuroscience, among others, discrete mathematics have a strong development. Indeed, there are situations where it is better to raise the discrete input and not the continuous model, as they are a better reflection. We must not forget that passing the mathematical model to the numerical model and computational solution is being a discretization.

On the other hand, apart from the binary logic (0-1, black-white) has emerged fuzzy logic that can take truth values in the range [0,1] (the full range of gray), and the theory of fractals offer new possibilities to mathematical modeling.

Companies' financial information is limited ultimately to relevant information about the situation and development of the economic and financial situation of the company. It can be considered, on the other hand, as a subsystem of "integrated system management information", ie, a system that connects all kinds of information produced in the company, such as statistics, forecasting, billing, payroll, accounting, in a joint treatment.

In this regard, the progress made in recent years are spectacular thanks to computer resources and teleinformáticos existing today in day.

IMPORTANCE OF MODELING OF FINANCIAL INFORMATION

The accounting analysis is an integrated information system within the overall financial reporting system - that is economic accounting. It aims to reading and analyzing information and interpreting the results thereof.

This financial information is not made to be hidden, but to be communicated to different recipients (owners or partners Business, Public Administration, Financial Institutions, etc.)

The field of accounting knowledge and financial information also covers those techniques pursuing information verification (audit) carried out by independent professionals of the company.

Accounting and financial reporting includes within its knowledge and techniques, analysis and interpretation of the information prepared. A bank, for example, not only reads the accounting documentation that provides the applicant for a loan, but interprets it to see whether it will be able to return at maturity.

The structure of financial information is related to the goal, to achieve through the operation of various functions within the financial information:

- Information processing
- Communication
- Verification (accounting audit)
- Analysis and interpretation by the user

It is specifically the task of preparing the coating to be more important since it will involve the result of the remaining functions. During the preparation of accounting information it will be necessary to observe parameters so that this meets the characteristics required:

A) Recognize and identify facts that define accounting information.

B) a measurement in monetary units of such accounting events.

C) a synthesis of information: This section accounting reports among which the most representative are the balance sheet will reflect the statement of assets, liabilities and net of the organization at a given time are made, and the Profit that displays the set of income and expenses over a period and according to these economic result for the period and losses.

The company is oriented to satisfy market and generate wealth so that financial information should be oriented to the market and generate wealth must report from a financial standpoint on:

- Wealth creation
- The return on investment of the partners or shareholders
- Generating cash flow

Wealth generation is known and measured through the income statement. The return on investment of the partners or shareholders is obtained from the information contained in the statements of financial position, performance and changes in stockholders' equity.

The financial information that has value is one that has an impact on our actions, answering our questions, which is based on decision-making as a competitive weapon to help the company drive forward to capture market improve profitability, improve efficiency, etc.

The significance of the information is based on the ability to represent symbolically in words (concepts) and quantities results of operations and financial condition.

Financial information must first meet internal needs and should be aimed at managers of the company so they can manage, make decisions and achieve goals. Secondly must meet the general needs financial, addressed to stakeholders who are not directly involved with management, and finally must cover fiscal needs. The information must come only from a single source which is the general accounting system.

Financial information is essential in making strategic and tactical decisions so their system and intelligent preparation is important so that it can be used as a competitive tool. It should be seen as a product, business support, which provides added value for decision making and should be effective in terms of timeliness.

Information should be ordered for each particular company oriented markets, customers and products with which it participates. This way one can measure the company in its economic environment.

Financial information is an important tool for analyzing the results of the company source but it is also necessary to know the competitive environment in which the company operates, evaluate the marketing system, quality of products, price, quality of staff, etc. Customers are a power group that impose conditions such as a minimum level of service, product location. The price has a pressure in competition and profitability of the company. So the departmental and cost center information is needed so that those responsible can monitor and manage their maximizing their results area.

To mediate making a decision, one should plan what one want to do in order to be able to give answers to the questions why and how. If these questions are answered, decisions taken were appropriate, since taking a good decision all those involved in the national economy we will benefit (government, employers, employees, etc.)

That said, we must assert ourselves of proper planning. For this reason, it is important to note that planning is a universal activity. Although their content may vary at different levels in the structure of the organization, there are certain steps which always occur that is planned. These steps are similar to those that always hit a process that involves making decisions;

then, in a sense, make plans is just a special case of decision-making with a strong orientation towards the future. For this situation it implies anticipate future activities concerning proposed decisions and the future outcome of present decisions.

The fact that we make a decision to a problem or action plans should be sought in accordance with the problems and solutions of tomorrow. Forecasts and research conducted carefully are the keys to making an excellent decision because selecting a decision, it must be based on criteria to be valid in the present and in the future.

Good decision-making must be based on facts and not on vague and general emotions. The activities are based on situations that dictate the facts, which can avoid obstacles, and should not be avoided, are recognized as existing and included in the plans, along with the respective provisions to deal with them.

The facts relevant to the situation considered relate to the experience and knowledge of who uses it.

The office of executive decision-making involves a certain degree of financial risk in a strategic framework defined by the company itself and a predetermined economic and social environment. The suitability of these decisions will depend on the preparation, experience, personality and information held by the management.

In decision-making experience is a key element because decisions must be made on a highly complex reality because of the huge number of variables that come into play.

The accumulation of experience is long and costly. If we consider that the more one learn is as a result of mistakes, reaching a high level of experience in the business world can have a terribly high cost. The immediate consequence is that the whole experience to be gained without the effects that might result from an erroneous decision or simply a non-optimal decision will be well received and cheaper, whatever the cost.

It is hardly possible to imagine a field of greater importance for human than decision making. We have a problem when we do not know how to follow. Once we have a problem, take a decision (including doing nothing). We chose an alternative that seems to us rational enough to allow us more or less to maximize the expected value after our resolute action. Quietly issued a control plan that guides us in making decisions, including decisions related to modify that plan control.

Finance, long considered as part of the economy, emerged as an independent field of study early last century.

Originally they were associated only with documents, institutions and procedural aspects of capital markets. With the 1920s, technological innovations and new industries brought about the need for more funds, promoting the study of finance to highlight liquidity and financing companies. Attention focused instead on internal management in external operation. By the end of the decade it intensified interest in the securities, especially common stock, making the investment banker a figure of special importance for the study of corporate finance period.

Depression in 1930s forced to focus the study of finance in the defensive aspects of survival, preservation of liquidity, bankruptcies, liquidations and reorganizations. Conservative tendencies dominated, giving more importance to the company to maintain a solid financial structure. Abuses with debt, especially debt related to the holding utilities, were exposed to many businesses collapsed. These failures, along with the fraudulent way they were treated numerous investors, grown demand for regulations. They increased the financial information that companies should disclose, and this in turn caused the financial analysis was broader,

During the 40s finance they followed the traditional approach that had developed during the previous decades. The company is analyzed from the point of view of someone outside it, as it could be an investor, but without putting emphasis on decision-making.

In the mid-1950s they became important capital budget and considerations relating thereto. New methods and techniques for selecting capital investment projects led to a framework for the efficient distribution of capital within the company. The financial manager now had charge of the total funds allocated to assets and capital allocation to individual assets on the basis of an appropriate acceptance criteria and objective.

Subsequently appeared complex information systems applied to finance, which enabled the realization of more disciplined and profitable financial analysis. Was deeply affected electronic means used by companies to conduct their banking, pay your bills, collect the money they are owed, transfer cash, determining financial strategies, manage currency risk, etc. valuation models for use in making financial decisions were devised.

The highlight of the 1960s was the development of portfolio theory (Markowitz - 1960, later perfected by Sharpe, Lintner, Fama and others) and their subsequent application to financial management. This theory explains that the risk of an individual asset should not be judged on the basis of possible deviations from expected performance, but in relation to their marginal to the overall risk of a portfolio of assets contribution. Depending on the degree of correlation of this asset with others that make up the portfolio, the assets will be more or less risky.

In the 1970s they began to apply the pricing model of capital assets Sharpe for valuing financial assets. The model of the risk implied that the company did not matter to investors of a company because it could dilute the portfolios of shares in its possession. Also caused it to focus even more attention to market imperfections when selecting assets by the company, financing and dividends are judged.

Also during this decade, Black and Scholes model formulated pricing options for assessing relative financial rights. The existence of an options market allows the investor to establish a protected position and risk by purchasing shares and, at the same time establishing stock options. Efficient financial markets in performance caused by a position of this type must be a risk-free rate. If this is true, it would be possible to establish exact formulas for valuing different types of options.

In the 1980s, there has been significant progress in the valuation of companies in a world where uncertainty reigns. It has placed a growing attention to the effect of market imperfections have on the value. Economic information allows for a better understanding of market behavior that have financial documents. The notion of an incomplete market, where the wishes of investors of particular types of values are not met, puts the company in the role of carrying out the marketing of special types of financial claims.

In the 90s, finance has had a vital and strategic role in business. CFO has become an active part: the generation of wealth. To determine whether generates wealth must know who provide the capital that the company requires for profit. This becomes the basis of opportunity cost, for which the product will be judged, investment and operating decisions. As global financial markets are increasingly integrated, the finance manager should look for the best price and often national borders currencies and other barreras-

External factors influence every day in the financial manager: deregulation of financial services, competition among capital providers and financial service providers, volatility of interest rates and inflation variability of exchange rates currency reforms tax, global economic uncertainty, external financing problems, speculative excesses and ethical problems of certain financial businesses.

In short, the study of finance descriptive study evolved from its early days until the current normative theories rigorous analysis. They have ceased to be a fundamentally concerned about raising funds to cover the field asset management, capital allocation and valuation of companies in a global market.

Differential equations used for modeling

A differential equation is an equation containing derivative of an unknown function. If this function is real variable, ordinary differential equation (ODE) says, and if the unknown function is multivariable equation is partial derivatives (EDP).

The general form of an ODE will represent

$$F(x, y, y', y'',..., and(N))=0, y=f(x) \qquad (1.1)$$

The order of an ODE is the order of the highest derivative.

Form (1.1) is a relationship between n + 2 variables, we assume that it is always possible to solve for y (n) in terms of the other variables, giving the normal way:

$$and(n)=f(x, y, y', y'',..., and(n-1)) \qquad (1.2)$$

ODEs are linear if F is a linear function of the variables and (). Then they can take the form

$$n(x)\ and(n)+ an-1(x)\ and(n-1)+....+ a0(x)\ and(n)= g(x)$$

Whether the function g (x) is equal to zero speaks of a homogeneous equation and otherwise is the inhomogeneous equation. Recall that given a differential equation have the following problems:

- Existence of the solution
- Uniqueness of the solution
- Calculation of the solution (analytical methods) or a "proxy" it (numerical methods)

The variable differential equations here noted x, may be the time noted t, in which case the event is a dynamic phenomenon. In turn, it may be necessary to determine whether the behavior is periodic. Moreover, the equations can acknowledge a time shift giving rise to equations with delay.

If the equation depends only on spatial variables speaking of autonomous equations. Besides equation, in practice must impose additional conditions that may be starting or edge (or boundary).

In most cases it is impossible to represent the solution analytically (using formulas). The same equation can give information on the characteristics of the solution. What matters is the long-term qualitative behavior; for example for a population it is important to determine if they maintain or tend to disappear.

Some models may require not one but several equations, ie a system.

An interpretation of the derivative as its sign corresponds to the growth of the function in the positive case, or decrease if the sign is negative. This enables the first order differential equations to describe chemical reactions of decomposition, populations or economic growth.

Population models and growth models

As the number of individuals of a population is whole, it seems impossible in principle to use differential equations to describe their growth. So we must assume that large populations change continuously differentiable with respect to time.

We denote p (t) of a given population at the time t, and p denotes the difference between birth rates and mortality population. We also assume that the population is p0 at time t0.

a) Act Malthus growth

Under the assumption that the population is isolated, ie there is no emigration or immigration, the rate of population change is proportional to the population at that time.

$$\frac{dp(t)}{dt} = r(t, p)p(t) \qquad (1.3)$$

In the simplest model, Malthus believed that r (t, p) is equal to a constant, ie r does not depend on time or population. In this case the solution of initial value problem is:

$$p(t) = p_0 e^{a(t-t_0)} \qquad (1.4)$$

The growth is exponential over time. This model is unreasonable for the human population, but is well suited to small rodent microtus arvallis pal.

b) Act logistics population

Equation (1.3) does not realize competition develop together individuals of the same population for the limited living space, natural resources and food available.

One have to add a name given by -bp2, where b is a constant competitive term, as the statistical average number of encounters per unit time is proportional to p2.

Then we have to solve the initial value problem:

$$\frac{dp}{dt} = ap - bp^2 , \qquad p(t_0) = p_0 \qquad (1.5)$$

This equation is known as the logistic law of population growth and b coefficients will be vital coefficients it says. The solution of (1.5) by separation of variables is

$$p(t) = \frac{ap_0}{bp_0 + (a - bp_0)e^{-a(t-t_0)}} \qquad (1.6)$$

Let's examine the type of population that predicts. When $t \to \infty$ we have that $p(t) \to \frac{ap_0}{bp_0} = \frac{a}{b}$

That is, regardless of the initial value the population always tends to the limit value a / b. An analysis of the second derivative of p (t) we conclude that $\frac{dp}{dt}$ is increased if and $\frac{dp}{dt}$ is decreasing if .

Fig. 1.1 shows the logistic curve (at S) if p0 <a / 2b. From the way it is concluded that the population before reaching half its time limit value is a period of accelerated growth. After this point, the growth rate decreases to zero.

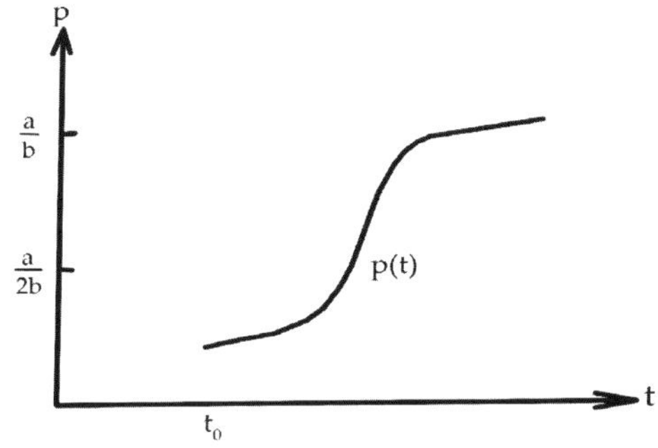

Figure 1.1: The logistic curve

A confirmation of this logistic law was held experimenting with the protozoan Paramecium caudurum. Observations are the following:

a) If signs of a and b are changed in equation (1.5) is shown in Fig. 1.1 that from a threshold ("critical population") can go into extinction or grow without limit, depending on the values of a and b.

b) could enrich the model considering the one hand, a critical population, so that starting from less than this critical value population go to extinction, and secondly, a maximum population, so that higher values growth is unlimited.

In the case of two populations, one will have to propose a system of two differential equations.

Models of economic growth

Both the derivative and the integral of a function and its endpoints are interpretations economic. There are situations involving differential equations or difference equations, depending on whether a continuous or discrete situation.

The Domar growth model assumptions are as follows:

Any change in the flow rate per year investment I (t) affect aggregate demand and production capacity.

The effect of a change demand operated in I (t) instantaneously acts through a multiplier effect. An increase in I (t) will increase the flow rate of income per year Y (t) for k = 1 / s increase of I (t) multiple. Here s represents the marginal propensity to consume. So $\frac{dY}{dt} = \frac{dI}{dt}\frac{1}{s}$

The effect investment capacity must be measured by the change in the potential production rate that the economy is capable of generating. Yes κ represents the ability or the potential annual production and capital flow K, we can assume that the relationship κ / K = ρ is constant.

Therefore, $\frac{d\kappa}{dt} = \rho\frac{dK}{dt} = \rho I$

Domar model assumes that the equilibrium is defined by a situation where productive capacity is fully utilized; this is Y = κ , Which allows us to write

$$\frac{dY}{dt} = \frac{d\kappa}{dt}$$

The question is to determine the time path of investment I (t) that satisfies this condition of equilibrium in any period.

Combining the above relations we get to the differential equation

$$\frac{dI}{dt}\frac{1}{s} = \rho I$$
,

assuming an initial investment I (0) is the solution of the equation

$$I(t) = I(0)e^{\rho st}$$

The result is exponential and therefore disturbing from the economic point of view. By modifying the hypotheses more flexible models such as Solow are obtained.

CHAPTER 2

MATHEMATICAL MODELING OF LOAN REPAYMENT

Ivan Stanimirović

In the business world, it is quite common contract is a debt to pay it in the medium and long term. Considering the fact that the nominal value of each payment consists of a mixture of interest payments and repayment of principal, one can use various methods to establish how to settle it a debt. For illustrative purposes, we remind that the crisis that has been going through the Available Financial System, together with the stabilization of the economy, led to a number of alternative financing, consortia and cooperatives in real estate. Such practical situations constitute the applicability of the subject matter hereof; greatly in the systems used most frequently.

Amortization systems are basically designed to loans and long-term financing operations involving periodic disbursements of principal and financial charges.

A key feature of amortization systems to be studied is the exclusive use of compound interest criteria, focusing exclusively the interest on the outstanding balance (amount) determined in the previous period.

Initial concepts

Amortization is the payment of principal or borrowed capital that is donc, usually, periodically and successively during the period of funding.

Interest is the cost of capital taken from the aspect of the borrower and the return on capital invested in the aspect of the lender.

Installment is the payment of amortization plus interest on the outstanding balance immediately prior to the period for the provision. The interest rate can be pre- or post-fixed, depending on contract term. It is understood as a fixed one whose expected future inflation rate is already incorporated into the rate, while in the post-fixed there is a need to ascertain to devaluation due to inflation, offset it through indexation.

Balance due or debt status is the amount due in period, immediately after the completion of the payment for this period.

AMORTIZATION SYSTEM CONSTANT (ASC)

This type of system, as its name suggests, is the steady repayment of principal during the term of financing. The provision to be paid will be decreasing to the extent that the interest will focus on a balance due dwindling. The amount of depreciation is calculated by dividing the initial capital and the number of installments to be paid.

To mount the sheet, we first calculate the amortization of the value:

$$A = \frac{P}{n}$$

where A is depreciation, P is principal and n is the number of benefits

- Interest is earned on the previous balance due to the result of calculation period.
- The provision is the sum of amortization to interest calculated for the period.
- The outstanding balance is the sum of the previous balance interest;
- The current balance is the difference between the outstanding balance and the provision.

FRENCH SYSTEM OF PRICE AMORTIZATION - (FSA)

This system provides, unlike SAC, the benefits are equal and successive throughout the term of repayment. It is important to note that, as the services are performed, the outstanding balance is decreased implying thus a concomitant decrease in calculated interest for the periodunder review. However, in function to keep the uniformity with respect to the value providing the amortization increases to compensate for the reduction of interest.

The provision of value is calculated from the PRESENT VALUE OF FACTOR FOR MULTIPLE OPERATING using the formula:

$$R = P.\frac{(1+i)^n.1}{(1+i)^n - 1},$$

where R determines the value of benefits, i denotes the interest rate and n is the number of time periods.

COMPARATIVE ANALYSIS OF ASC AND FSA METHODS

To compare the two methods presented here, we will make a joint study from a following hypothetical situation:

Principal: $ 15,000.00

Interest Rate: 10% after

Number of time periods: 10

1 Calculation of the amortization for the SAC

$$A = \frac{P}{n} \qquad A = \frac{15000}{10}$$

2. Calculation of benefit to the SAF

$$R = P.\frac{(1+i)^n.1}{(1+i)^n - 1} \qquad R = 15000.\frac{(1+0{,}10)^{10}.0{,}10}{(1+0{,}10)^{10} - 1}$$

R = 2442

As we can see from the analysis of spreadsheets, the benefits of the SAC are larger than those of the SAF at the beginning of the period, getting smaller in the end. Of course, there is a moment at which the equality of payments. We can calculate this moment through the following reasoning:

Since j = P i. n (interest) we have that

$$A = \frac{P}{n}$$ (Amortization)

R j = A + (provision)

So, it is valid that

$$R1 = \frac{P}{n} + P.i.1$$

Paying attention to the spreadsheet SAC noted, from the 2nd period, the benefits have arithmetically decreasing values, then we can expose in mathematical terms:

R2 = A + {j - [(1 / Pn.).i]}

$$R2 = \frac{P}{n} + \frac{P.i(n-1)}{n}$$

Indeed, we have that

$$\text{R2} = \frac{15000}{10} + \{\frac{15000.0,10.(10-1)}{10}\} = R2 = 2850$$

We can generalize this formula as follows:

$$Rk = \frac{P}{n} + \{\frac{P.i.[n-(k-1)}{n}\}\}$$

Where K represents the analysis period.

AMERICAN SYSTEM OF AMORTIZATION (ASA)

This system is characterized by the fact that the return of the principal being made at once at the end of the amortization period. Intermediate depreciation is not provided for the duration of payments, with interest paid periodically. Due to the financial impact that the system itself conceives aimed at single debt payment, it is necessary to set up a sinking fund, which will fund that amount. It is important to note that this fund will be made concurrently with payment of interest on the principal by using several operations in capital accumulation factor.

In short, this amortization schedule the principal is repaid through a single payment at the end of the operation. Interest may be paid periodically (most common) or capitalized and paid together with the principal at the end of the agreed term.

Mathematically we have

$$R = Sp\frac{i}{(1+i)^n - 1}$$

APPLICATION IN NET PRESENT VALUE

In a financial transaction of investment or financing, there are several situations that interfere with our decision about choosing one of several possible alternatives. In general, we have the knowledge of the market rate, also known as the Market Attractiveness rate and desire to know the real rate of interest of the operation, to be able to make a decision.

There are two important mathematical objects that are used in the analysis of financial transaction Investment and Financing: Net Present Value (NPV) and Internal Rate of Return (IRR).

Every action done in the financial domain necessarily implies the pursuit of optimizing the cash flow generated by this action, that is, income and expenditure flows. Using the basic idea of financial mathematics that allows, through the capitalization and capitalization, comparison of capital in time, we will prove one of the methods used in choosing the alternative that provides greater profitability and lower cost.

Net Present Value (NPV)

The method of Net Present Value (NPV) aims to value in terms of present value impact of future events associated with a project or investment alternative, ie measures the present value of cash flows generated by the project over its life useful. In the absence of capital restriction, it is argued that this criterion leads to great choice as it maximizes the value of the company.

The purpose of the NPV is to find projects or investment alternatives that are worth more to the sponsors of that cost - projects that have a positive NPV. Its calculation reflects the preferences between present consumption and future consumption and the uncertainty associated with future cash flows. The process through which the cash flows are adjusted to these factors is called off, and the magnitude of these factors is reflected in the discount rate used (cost of capital). The discount process converts the future cash flows to present values because of different times cash flows cannot be compared or aggregated until they are placed in the same time.

The calculation of NPV is the difference between the Future flows from the duly updated initial capital through a rate equal to the opportunity cost and this Capital. The decision criterion implicit this parameter is as follows:

1. If NPV is greater or equal to zero means that the flow will be economically feasible, meaning the project conceive a greater or return at worst, equal to the required return and, therefore, accept the project; otherwise it is being rejected.
2. If we have several possible flows over the same capital, will choose the one that produces GREATER net difference, as is one that provides GREATER financial profitability.

Mathematically we have

$$NPV = [\frac{N}{(1+i)^1} + \frac{N}{(1+i)^2} + \frac{N}{(1+i)^3} + \ldots\ldots \frac{N}{(1+i)^n} - P]$$

where NPV is the net present value, N is Nominal Value, P is capital and i determines the interest rate.

CHAPTER 3

MATHEMATICAL MODELS IN E-BUSINESS

Ivan Stanimirović

A specific company can use several business ranges define, particularly those coming from the industrial economy, and which are transforming the internet. These also remain for some time, traditional ways of running business. On the other hand, there are new companies in the information economy, using only varieties of businesses that are peculiar to it; for example, information portals, online auctions, electronic encyclopedias, search services, etc.

Very popular first differentiation in the art and literature is popularly distinguish cases where a business sells Internet to the final consumer (Business to Consumer: B2C) and those in which a business sells another business (Business to Business: B2B). It is clear that the characteristics of both types of businesses are very different and the challenges they face. Indeed, in many cases it is simply B2C Internet commerce with all that this implies in terms of selection and customization of products, marketing, customer care and prices- with a major challenge: to provide a product that replaces the advantages over traditional offer or does not exist in this for example, webucation replacing education / face-to-face or electronic auctions of products that do not end in the traditional-options or offer the same products of traditional training providers, but can deliver significantly more convenient conditions for example, books, computers, videos, etc. on the Internet. However, this trade is not to just copy Internet features traditional trade. Due to the massive capacity of the Internet for many people to access a site, those who are successful in selling certain product lines can constantly expand their offerings, to have the attention of a critical mass of customers.

This may involve the independence of an e-commerce site physical supply of product, because it can sell or broker for others. This trend is already present on Amazon.com that is expanding from books in multiple directions, with some products provided directly by the company and others sold on behalf of other providers. It also goes in the same direction that portals such as Yahoo, which were essentially pure content buscadores-, have evolved into offering other physical products. In other words, the evolution of B2C is from direct sales of products to intermediaries, for which it is prerequisite to have a site that captures the attention of many customers, by giving a single value; for example, a wide range of product choices with information and support that allows the customer to select efficiently. This trend is already present on Amazon.com that is expanding from books in multiple directions, with some products provided directly by the company and others sold on behalf of other providers. It also goes in the same direction that portals such as Yahoo, which were essentially pure content buscadores-, have evolved into offering other physical products. In other words, the evolution of B2C is from direct sales of products to intermediaries, for which it is prerequisite to have a site that captures the attention of many customers, by giving a single value; for example, a wide range of product choices with information and support that allows the customer to select efficiently.

This trend is already present on Amazon.com that is expanding from books in multiple directions, with some products provided directly by the company and others sold on behalf of other providers. It also goes in the same direction that portals such as Yahoo, which were essentially pure content buscadores-, have evolved into offering other physical products.

In other words, the evolution of B2C is from direct sales of products to intermediaries, for which it is prerequisite to have a site that captures the attention of many customers, by giving a single value; for example, a wide range of product choices with information and support that allows the customer to select efficiently.

With some products provided directly by the company and others sold on behalf of other providers. It also goes in the same direction that portals such as Yahoo, which were essentially pure content buscadores-, have evolved into offering other physical products. In other words, the evolution of B2C is from direct sales of products to intermediaries, for which it is prerequisite to have a site that captures the attention of many customers, by giving a single value; for example, a wide range of product choices with information and support that allows the customer to select efficiently.

They have evolved to offer physical products of others. In other words, the evolution of B2C is from direct sales of products to intermediaries, for which it is prerequisite to have a site that captures the attention of many customers, by giving a single value; for example, a wide range of product choices with information and support that allows the customer to select efficiently. They have evolved to offer physical products of others. In other words, the evolution of B2C is from direct sales of products to intermediaries, for which it is prerequisite to have a site that captures the attention of many customers, by giving a single value; for example, a wide range of product choices with information and support that allows the customer to select efficiently.

On the other hand, the B2B focuses on providing unique value to the participants in the exchange. In a situation in which there is intermediary between two companies for example, a market (exchange) mail where supply and demand for certain products- is traded, the value for participants comes from the great transparency and market efficiency [18]. Now, when the relationship is directly between companies, the value one can capture the offeror and / or applicant. One case is that of an offeror who has a site that allows customers to order their companies their products online. This provides value for itself and the customer by reducing transaction costs and also to itself for a possible increase in demand for service improvement. On this last case, it is possible to give the customer value-added services such as inventory management or logistically, which also creates value for the offeror wholesale customer loyalty and possible increase in demand. Another case of a direct relationship is one in which the plaintiff which obviously should buy large and attractive volumes-has a website through which publicizes its demand and accept offers. In this case, the power of the plaintiff makes the value is mainly for him, for better prices induced by the expanded competition from many suppliers. A variation of this scheme is a consortium of several plaintiffs, gathering their needs, increase their power to get better prices, which also creates value for the offeror wholesale customer loyalty and possible increase in demand.

Another less popular classification scheme, but as important that the newly introduced, is based on the type of products traded on the relationship e-Business. The products traded over the Internet can be physical or digital. Physical products are those involving a material flow provider customer (individual or company): books, videos, steel, cars, money, equipment repairs, etc. Digital generated only are the electronic flow of information: -Bookings digital services and travel ticketing, insurance, consultations dictionaries and encyclopedias, etc., digitized music, «new» -búsquedas information, consolidated product offering content, buying tips, search for better offers, education and training electronically, electronic consulting,

In Table 2.1, the intersection of the above classifications shown, which gives rise to a type of e-business, where some specific businesses also exemplified within each category. The criteria were used in this classification are as follows:

e-Tailing: physical products sold to the final consumer supported on a website; for example, sale of books, videos, CDs, DVDs, cars, etc. It may be the distribution of a product that makes other direct sale of a product manufactured by the same company or a sale brokered by a third party; for example, an electronic auction physical products oriented consumer.

e-Commerce: Sales of digital products or services, such as reservations and tickets and tickets, insurance, CD digitized, software, various -News content, advice, searches, financial information, etc, e-Learning and employment services. It may be mediated, where the product or service provided by a third party; direct, in which the same company sells and produces the product or service; or own or other content.

e-Sales: Typical that a company selling their product to other companies, based on Internet. The product can be physical or intangible, such as consulting, legal, medical, etc.

e-procurement: Typical supply by an enterprise of products or services required by a website. It can be a product or bephysical defect for example, spare parts or repairing a team- or intangible for example, development of computer applications, legal services, etc.

e-Market: New way exchange between companies through an electronic market media supply and demand, administered by a third party that guarantees transparency and efficiency. It may be by physical or intangible products or services.

Table 3.1: Types of e-Business

<table>
<tr><td colspan="3" rowspan="2">RELATIONSHIP
BETWEEN
PARTICIPANTS
PHYSICAL
DIGITAL</td><td colspan="2">PRODUCTS</td></tr>
<tr><td></td><td></td></tr>
<tr><td rowspan="6">B2C</td><td rowspan="4">It PROVIDED AND EXISTING PRODUCT</td><td rowspan="2">Generated by another</td><td>e-Tailing distribution</td><td>e-Commerce intermediated</td></tr>
<tr><td>Amazon
iQvc [38]
CarsDirect [38]</td><td>Ticketmaster
Napster
Pricelined [14]</td></tr>
<tr><td rowspan="2">OWN</td><td>e-Tailing direct</td><td>Direct e-Commerce</td></tr>
<tr><td>Dell
Barnes & Noble
Lands' End [38]</td><td>Thrive Online
Merriam-Webster
Britannica</td></tr>
<tr><td colspan="2" rowspan="2">PRODUCT IS PROVIDED "NEW"
e-Bay</td><td>e-tailing intermediated</td><td>e-commerce content</td></tr>
<tr><td>Google
yahoo
MySimon [41]
Expedia [14]
CareerPath [13]</td><td></td></tr>
</table>

		e-Sales physical	**e-Sales intangible**
B2B	**DIRECT CONTROL** **OFFERER** Cisco Sigma-Aldrich [34]	Andersen Consulting	
		e-procurement physical	**e-procurement intangible**
	PLAINTIFF WITH DIRECT CONTROL Covisint [35]		
		e-Market physical	**e-Market intangibles**
	intermediated Enron [16] Ariba [23] ChemConnect [11] iPlanet [33]		

Like any classification, this is an idealization based on pure types. Obviously, there is a possibility, and it is given in practice that an e-business mix different types. For example, Yahoo, as already mentioned, which began as a pure content provider, is now involved in selling physical products, sticking out an additional party to its huge customer base [39]. In addition, the inner workings of companies is hidden in this classification, since all processes satisfying the requirements for products or services transacted in business are not explicit. However, we assume that the design of a business should not only include interaction with other agents but also all the back-office makes possible such a relationship where Internet technology is equally applicable.

It is shown in Table 3.1 that different businesses have, by their very variable characteristics, challenges change from type to type. For example, in a first rough approximation, businesses that handle physical products have the fundamental problems of logistics. By contrast, digital products move efficiently on the network and the challenge is to have an insurmountable site in terms of attractiveness, usefulness and efficiency.

CHAPTER 4

ECONOMIC CONCEPTS AND THEIR APPLICATION IN E-BUSINESS

Ivan Stanimirović

Let us examine a series of economic concepts that come from the industrial economy, but, properly applied, are very relevant in explaining the operation of the e-Business and provide guidelines for design.

MICROECONOMIC THEORY OF THE FIRM

This theory focuses on characterizing the short production function of a company, in order to provide a framework for production decisions. For this defines the following total cost function:

TC (q) = FC + TVC (q) (1)

Where:

$TC(q)$: Total cost of production, which depends on the level of output q

FC : fixed production costs, which are incurred regardless of the level of production.

$TVC(q)$: Total variable cost of production.

This cost function is short term, which assumes that the machinery and plant generally remain unchanged. Examples of fixed costs are personnel costs unchanged-say, managers and administrative personnel with the level of production and the cost of leasing a non-own production facility. Examples of variable costs are the costs of production inputs and energy use to produce the product.

The function (1) has shown graphically in Figure 4.1. From it are defined:

Average variable cost function:

$$AVC(q)=\frac{TVC(q)}{q}$$

Average total cost function:

$$ATC(q)=\frac{TC(q)}{q}$$

Marginal cost function:

$$MC(q)=\frac{\partial TC(q)}{\partial q}$$

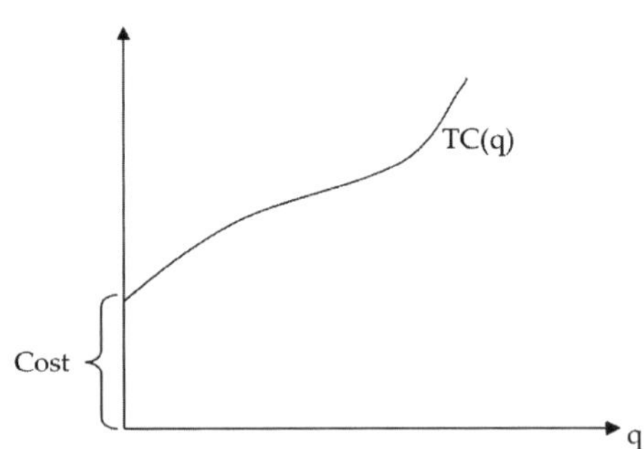

Figure 4.1: Total cost function

These costs can be represented graphically as shown in Figure 4.2.

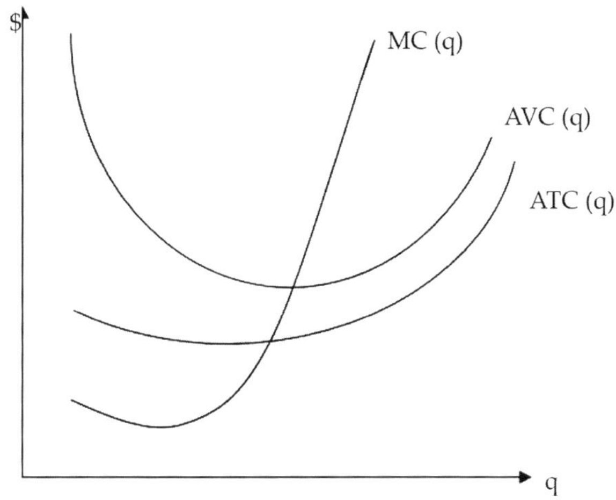

Figure 4.2: Average cost functions and marginal

The marginal cost is an important interpretation, which is to reflect the cost of the next unit to produce, when the current production level is q.

Of course, all the above approach is highly simplified, since, among other things, assumes aggregate production -no considers each particular product or the mixture of them and continuity of production. However, as we shall see, can draw some interesting conclusions regarding the functioning of a company. In particular, one can set the conditions for optimum production level of a company as follows:

Defining p(Q) as the total utility of a business, we have:

$$\pi(Q) = pq - CT(q) \qquad (2)$$

where p is the price at which the product (infinitely elastic in a competitive market) demand.

Differentiating (2) with respect AQE equating to zero to obtain the utility-maximizing q is reached:

CM (q) = p

In other words, the company should produce the level at which the marginal cost of production equals the short-term market price of the product.

One can also get the production cost function long term, which can vary the sizes of plant and equipment for different production levels. This is represented by the average long-term cost (CPLP), as shown in the curve in Figure 4.3. The rationality of curve is that as q increases production level, larger plants producing economies of scale and reduce the average cost are used. But, at a certain level they are produced diminishing returns-and scale, for example, costs increase coordination and control more than proporcionalmente- with increasing the scale of operations.

Once examined, the behavior of the above costs and market structures in companies that are in e-Business, discriminating according to the types defined in the previous section.

Consider, first, the companies conducting e-Business digital products. These companies are typically the economics of information -that say, are born with the development of Internet- although there are cases of traditional companies, following the use of the Internet as a channel of digital distribution, are transforming physical products to digital . Examples of the former are AOL and Yahoo, and second, Microsoft and producers and distributors of videos, DVDs and CDs.

These companies have high fixed costs and low variable cost very short term. Indeed, companies that provide digital content such as electronic encyclopedias, directories of various types, electronic maps, etc. incur considerable when creating the product and marketing approach, which is the fixed cost cost. The short-term variable cost of providing the product is almost negligible, since it involves mere reproduction of the digital and Internet distribution.

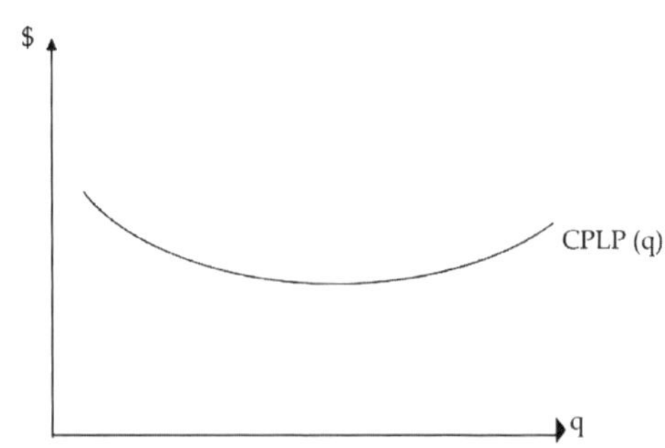

Figure 4.3: Total cost function of long-term production

On the other hand, companies that provide more akin to traditional digital products such as software, videos, music, etc. have the same characteristics. That is, the marginal cost of producing and distributing an additional copy online is virtually zero and, in some cases, is valid for original physical products such as software.

Companies with the above features may have little traditional behavior, but justified in economic theory presented. For example, many of these companies "give away" the information or products, which explains why the marginal cost of producing and distributing them is practically zero, and try to be financed by advertising. Alternatively, some of them, like Yahoo, have tried to enter the traditional business selling physical products where one can earn an income associated with each transaction.

In other words, the information on the Internet becomes a "commodity" who have rights except patents- tending the price at marginal cost, which is close to zero, which means it is very difficult to compete and get profitability. However, there are strategies that can be followed and discussed in Section 4, which have to do with differentiation and customization of products, and price discrimination.

Market structures which tend to these products are also features. Economies of scale prevailing there lead to a dominant company, with a very low cost and not necessarily the best product-, especially when it becomes a de facto standard and there protecting it through patents. The paradigmatic case that illustrates this trend is Microsoft, a company whose physical products, and even more, digital products over the Internet, have the characteristics of high fixed costs and low variable cost.

However, companies that traded physical products online have, in most cases, fixed costs and characteristics similar to traditional variables. For example, Amazon.com, Cisco, Dell and the like. Clearly, these companies have a significant variable cost, which is generated in the production or purchase, inventory and distribution of products. Therefore, behavior and market structure tend to be more traditional. However, the use of Internet allows, in some cases, innovation in terms of products and services, which gives them competitive advantages. For example, Using historical information online customer behavior to customize products and make targeted offers, as does a supermarket that sells Internet that suggests buying a basket of its users based on their historical acquisitions

[22] - is It is generating a combination of product and service that gives one competitive advantage. Similar is the case with Dell that is integrated into its supply chain, through them know their suppliers production plans for these products will deliver “just in time”, which changes the supplier-customer relationship. Also, direct, Internet, including Amazon.com and Dell with the consumer tends to eliminate the need for traditional distribution chain.

On the other hand, companies such as Internet auction e-Bay, Enron, ChemConnect and the like also change the market structure to position itself between supply and demand, which traditionally operated in direct contact.

As the long-term average costs, the impact of the Internet, both companies that sell digital products and physical, is to a significant decline due to the impact of these technology that improves daily with decreasing costs. In particular, the ability to handle large scales of operation that provide new technologies -at facilitate coordination and control- reinforce the effect of economies of scale and reduce long-term average costs as volume increases. For example, Amazon.com uses in an exceptional way Internet technology and process to meet their customers and, by adding new equipment and software, more powerful and economical, it is able to handle increasing volumes of customers with decreasing average costs. On the other hand, the logistics, Amazon.com also has a very good handle for the current scale. However, on this side, a scale of increasing operations could eventually given the complexity of managing many products in numerous wineries, diseconomies out-for coordination and control. But Amazon.com could use additional technology to overcome this problem, as robotized warehouses and production / purchase “on demand” products, feasible on a larger scale, which further reduce their costs. This, combined with other effects, as discussed later switching costs and externalities in networks- reinforce the trend of market concentration in a few companies. Given the complexity of managing many products in numerous wineries, diseconomies out-for coordination and control.

COST COORDINATION

The cost of coordination is one of the components of production costs presented in Section 3.1. Along with materials, labor and several others which, as noted, can be increased as it grows the size of a company, desconomías of scale due to management complexity inherent in large dimensions. This can lead to the long-term average cost increase. We will see here how this cost is generated, the factors that determine their magnitude and the options available to bring it to an optimum level. We establish also how Internet technology influences the magnitude and optimization thereof.

On the basis ofl management of any company is the problem of how to deal with relationships outside the company and internal activities and / or how to handle relationships between internal activities. This relationship management can also be conceptualized as coordination among interdependent activities.

Now, what is a relationship or dependency? We can say that there is a relationship when the actions of an activity affecting the behavior of another and / or vice versa. Such a relationship can be handled in different ways. The default is to manage trivial in the sense of letting act to the activity that affects another and that this act reactively, based on the results observed. For example, shares of buying products from a customer, affect sales activity; it can handle this relationship reactively, hoping to get the order and, at that time, act to try to satisfy it. Likewise, stock sales affect production; a reactive management of this relationship would watch the stock of finished goods and production has dropped to act when an estimated critical level.

In a reactive management, we can manage a relationship as anticipatory, through explicit communication between activities that affect the affected, to try to detect their actions or intentions of action before they occur. For example, sales prospecting expected by customers to plan their satisfaction in advance and communicating these projections to production, so plan oner actions production accordingly.

The above cases are extreme situations how to handle relationships between activities. We will see that, depending on the type of relationships, there are several alternatives, each with different economic consequences. For purposes of our frame of reference, classify relationships or dependencies on the following types [5]:

a) Sharing: What happens when several activities share a scarce resource, which creates the need to decide the allocation of such resources; for example, when several products share the same facilities and people in their production or operation, which is the case of other manufactured articles in batches, in a single installation, and various products in a bank.

b) Consumer (client) happens when an activity produces something that is used by another; for example, a credit approved by an executive who happens to be executed by another unit, or when a winery provides a raw material for use in production. These relationships may be between internal activities to the company or between internal and external. They can, in turn, subdivided into:

   i) Restrictions sequence: in the case where an activity *provider*It must be completed before the start of the consumer; for example, an insurance policy can not be issued before rates have been established to collect.

   ii) Transfer: corresponding to the physical movement of requiring the consumer to the supplier; which can signifICAR move from large equipment or materials to books. In the case of transfer of information, usually one talk about communication.

   iii) Usability: which states that what is provided must correspond to what the consumer needs; byexample, the product or service a company should be usable by those who acquire it, and the information it receives an activity should be adequate to perform a certain action.

c) Concurrency restrictions: indicating that two actividades must occur simultaneously or, conversely, they can not be simultaneous; for example, a meeting requires several people are present at the same time, and a vehicle can not be used and repaired at the same time.

d) Task-subtask: that is when several activities are part or component of a larger task, to fulfill a certain purpose; such a company is organized (divided) in sales, production and finance, to fulfill its purpose of generating a good or service, or a construction project is structured in a series of sequential activities or parallel, under a project manager.

Now, we see that these types of relationships are in different situations that will raise and then, for management or coordination, there are several alternative mechanisms we shall explain in each case.

Relationships can be handled with greater or lesser degree of coordination, understood this as the highest rated less than having the actors involved in the process, achieve the ultimate purpose of it. For example, if one are processing a customer order, the sales agent may or may not be aware, accepting an order, whether it will be able to meet promptly or not -Objective end of this process-; production and agent may be conscious or not there is a request to meet.

One can induce greater coordination in a process by various means. These means or coordination mechanisms procome from various disciplines, such as Organizational Theory [15, 31], Economy [49] Operations [4] Research Information Technology [6] and general administration.

They can be used first, *rules*which used routinely by agents induce a degree of coordination. For example, the sales agent -at accept a pedido- can have as a rule consult the winery and reserve stock management, production and agent can have as a rule observe the stock, and once it reaches a set level, replenish the stock with a new production order. Obviously, in this example a degree of coordination is induced between organizational functions under certain circumstances, achieves the goal of meeting the needs of customers. However, the rules may fail because they are essentially static although may be reapplied, as the conditions by modifying, and an abrupt change in conditions can make them obsolete.

For example, replenishing stock based on point system works well -according to inventory theory [4] - while no marked seasonality or significant trends; if these are not given condiciones, the result of the rule is lack of stock in some cases -and therefore customers- dissatisfaction or excess stock in others. Table 4.4 displays several examples of rules are shown to handle the types of relationships defined above.

When considering alternative rules for a given situation, one can also use other ideas, based on information technology, fundamentally change the historical way to manage the relationship. Thus, for manejar the shares relationship, one can get away from the traditional idea that such allocation should be a rule applied by a person, and use an automatic decision algorithm based on expert system technology, in which the allocation It makes the computer, based on production rules derived from the experience of human allocators. This is the case of lending by an expert system, which allocates scarce resource, money, credits to certain competing for the same. In the same line of thought, when a shared resource is assigned, one must know their availability, with the possibility of moving away from the traditional idea that one have to directly establish the availability using monitoring technology, in which the same resource, "says where is". For example, in Chuquicamata automatically it knows the position of a truck mining transport within it, using sensors that capture radio waves emitted by such truck. Based on this, a computer which receives signals from the sensors-knows where each truck and can apply an algorithm that determines the allocation of which are free, to a new use, automatically. Other technologies that allow this monitoring in other contexts are bar codes and pattern recognition.

For the management of the supplier-consumer relationship, technology type *workflow* One can automate predefined sequence and documents respect prerequisite, replacing physical book flow routes.

Table 4.4: Rules for different relations

Relationship	Examples of Rules
Shared resources	A priori allocation based on the type of task
Provider-consumer	Notice of termination by the supplier
Restriction sequence	predefined task sequencing
Transfer	Definition of intermediate stocks (buffer); rule just in time
usability	Standardizing flow between tasks
Concurrency restrictions	Scheduling algorithms shared calendars (groupware)
Task-subtask	Management by Objectives

For concurrency, there are technologies shared databases including groupware-, the accesor remote to central computers and videoconferencing, all of which break traditional ways to handle this relationship and make it possible to apply rules that require the concurrence of several people or act simultaneously on the same information base; that is, the information may be available in several places at the same time to support concurrency. People in separate locations in space may appear as if they were working together in one place; for example, a seller may apply, ground rules to meet orders to a customer, connected wirelessly and the central computers of a company Internet, and a regional marketer of apparel, met by videoconference with offices core of a company,

Information Technology can also support radical changes in the management of relation task-subtasks, by rules; for example, can be eliminated verification rules and control, applied by a supervisor in charge of a task, they tend to ensure compliance with the purpose of it, for automatic compliance with them, based on a computer support, at the time of realization of the subtask, which internalizes the rules.

A second possibility of coordination is the use of the hierarchy. This means we deliver agents to a supervisor definitionthe performance of them and the solution of problems arising in such action. For example, in the production activities of a company there is always a first-line supervisor that establishes what the orders or work to be undertaken by each of the dependents in the machines available are; In addition, this person states what to do when unforeseen emergencies and other problems arise. The hierarchy also be used -of obvious that the rules-way to handle the different types of relationships defined above. Thus, a supervisor, through explicit instructions, can allocate shared resources, manage supplier-consumer relationships, coordinate and define simultaneity subtasks to perform a task.

However, it is obvious that the overhead hierarchy and, in many cases exceeds the capacity of coordination of people; For example, if the case manager had recently given a lot of dependents. Consequently, in medium and large enterprises managed well, one tend to use the hierarchy, except when the rules fail.

Thus we have, then, a third type of coordination mechanism: planning. Here there is an explicit instance of coordination -not necessarily hierarchically authority establishing and dynamically advance the role of each of the activities involved in the process, or part of a process- so that a target is achieved explicitly stated. An example of such coordination would be integrated the operations of a business including sales, production and abastecimiento- where, by an explicit model of planning- for example one of Linear Programming or one packaged within a planning ERP II -, a sales plan, a detailed production plan and a plan of acquisition of inputs is integrally generated. Of course, planning to operate it requires agents to deliver information on forecasted sales,

Planning also solves the coordination of relations already explained, as exemplified in Table 4.5.

In addition, information technology can amplify the power of planning, for the purpose of increasing the quality of coordination. It is obvious that production planning can be based on such technologyas the ERP; Planning

Financial, in projection systems such results; project planning packages supported by CPM / PERT type; and strategic planning Decisional support systems which use econometric techniques from market research to economic evaluation models to determine courses of action. Also, any of the schemes to coordinate by planning requires the status of what is planned, in order to establish future courses of action, for which monitoring technology and projection type is required; for example, multidimensional databases or data warehousing.

As a final coordination mechanism exists what we call collaborative coordination. In it, the actors involved in a process coordinated among themselves, based on active communication and participative decision. This type of coordination is given in very complex situations where creative work predominates. For example, to coordinate a working group working in the new product development in a company. As in the above mechanisms, it may be helpful to manage the types of relationships presented above. For example, the shared resource allocation is left to a working group; supplier-consumer relationships can be managed by analysis and solution level agents operating in the style of Total Quality; simultaneity, for coordination of the group; and by delegation. An interesting case of this kind was the management of supplier-consumer in the example of Hallmark [5] relationship. In this company, for the development of new products, a working group was established, avoiding the strict sequentiality and emphasizing work in parallel, also obviating the design review by the executive, delegating the decision to the group. Isproduced to eliminating a lot of delays that were in the sequential process and allowed to reach the goal of developing new products in less than a year.

This approach to the development of new products is what is now called concurrent design. Now, the predominant technology to support this type of mechanism is that of groupware, without prejudice to other tools, such as software programming joint agendas, simple e-mail and the like, are also useful. Monitoring technology is equally important, because through that every person involved in a complex activity-for example, a project with multiple sequence dependencies, concurrency, and others know the status of the actions of others, one can set needs action, delays and other information, by group participants.

Table 4.5: Examples planning to coordinate relations

Relationship	Example Planning
Shared resources	Production Planning
	Financial planning
Provider-Consumer	
Restrictions sequence	Planning CPM/PERT
Transfer	Activity program
usability	Queries the user
Concurrency restrictions	Planning CPM/PERT
Task-subtask	Strategic planning
	budgets

Error! Bookmark not defined.

In a given case one can freely choose the level of coordination will be exercised and, even more, one can choose not coordinate, breaking the activities of the company completely independent groups. For example, one could divide a factory that manufactures various kinds of products in various factories, each with a single type of product, to avoid coordinate interrelationships involving the manufacture of various products in the same facility.

The question is, then, what level of coordination choose or, if we have a low level of coordination now, to what level is appropriate to increase it. Also a relevant question is cuI walk to divide a company into smaller separate pieces, to avoid coordination.

The answer has to do with the visible and hidden costs that induce a degree of coordination. The obvious costs associated with coordination are Aquellos on the means used to coordinate. The more coordination, more people and more time devoted to coordination, more information and more processing hardware, software and communications to support coordination. If this were the only cost would not be justified coordination. However, there is a cost that displayed little in practice, which is associated with the consequences of not coordinating. The coordinate bit implies that the resources of the organization are used in a far less than optimal. In particular, the so-called slack resources [15], which are resources allocated implicitly or explicitly to absorb the consequences of the lack of coordination appear.

Slack resources are of various types. We have, first, the inventories of both materials and products terminados- that exist within companies, because no attempt is or is not able to coordinate explicitly and precisely the needs of customers with plans production company, nor the manufacturing to procurement by shopping. In relation to this clearance, Western--and some Japanese firms have proven in practice that such coordination is possible with a mechanical (rule) the type just in time and eliminate, therefore inventories.

Another resource slack is lowering standards. Thus, for example, a company is formed with a less than perfect quality, unwilling or unable to coordinate las productive activities to eliminate failures; another accepts a lower use of installed capacity to 100%, with the inability to coordinate customer demand with capacity and production activities to fully use; and a final, admits low productivities -in relation to other companies comparable, given the impossibility of coordinating labor, to induce high productivity.

Finally, it is also resource slack giving a service level to less than optimalos customers in areas such as order fulfillment, delivery, processing in obtaining the product or service, etc. In this case, the gap is provided by the customer, always willing, or alternatively not have to accept the level of service that is provided. But if the customer leaves, temporarily or permanently, to obtain the product or service in other companies, the play is provided by one.

Then, choosing an appropriate level of coordination becomes a problem economic: it is necessary to balance the cost of coordination with the cost of the consequences of failing to coordinate; these costs move in the opposite direction to increase the degree of coordination, as shown in Figure 4.6. We understand the degree of coordination as induced by a combination of the above explained mechanisms. According to this diagram, there would be an optimal level of coordination, where the sum of the two curves in Figure 4.6 is minimally, but obviously it is very difficult to calculate in practice. However, this analysis provides a conceptual framework, at least, says what factors must be identified and try to evaluate, to decide whether or not to increase the level of coordination.

The experience of the leading companies in e-Business both digital and physical products- it can be seen that coordination is largely automated, primarily through rules and planning. So work Amazon.com, Dell, Emron, e-Bay and many others, and could not be otherwise, since all activities involved in meeting customer requirements must work speed Internet, to process large numbers of transactions. Internet technology makes the above possible, to be applied not only to the interaction with customers and suppliers, but also the -Logistics internal management, financial management, production / storage, etc. particularly in e-Business with physical products. Of course, simultaneously Internet, can be used a number of Information Technology, mentioned above, as workflow, groupware, ERP,

Interpreting the above in light of the graph in Figure 4.6, we have to enterprises e-Business leaders have a high degree of coordination-and assuming that achieved a reasonably optimal balance between the cost of coordination and not associated with coordinar- the coordination cost curve would be shifted to the right. In other words, the Internet technology allows greater coordination exercise at a relatively low cost, achieving a balance at a level above it.

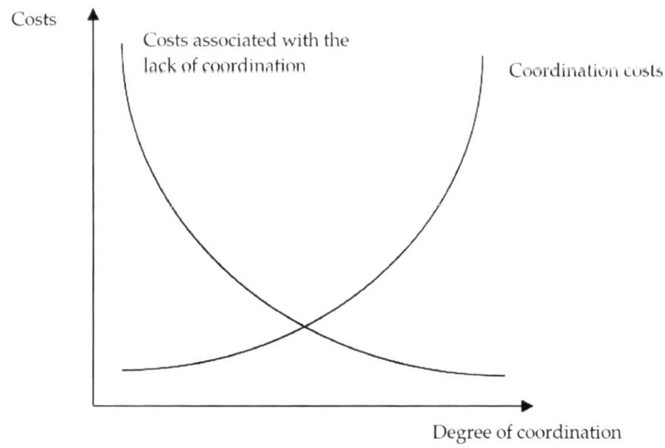

Figure 4.6: Balance coordination costs

This effect can be increased with the scale of operations in the long run, because they can use more powerful technologies, which are made feasible by having greater volume, lowering the cost by increasing coordination. This means a further shift of the cost curve to the right coordination and balance at a higher level. This effect is enhanced by intrinsic improvements in technology over time, as expressed in Moore's Law which states that the processing capacity of the basic components of computers doubles every 18 months decreasing cost.

In short, coordination with technology becomes cheaper with scale and time, which emphasizes its application, which is what we're seeing happening with companies e-Business leaders. Obviously, this contributes to lower average long-term costs, as the balances that move to the right in Figure 4.6 scale are given a level of total cost lower and lower.

TRANSACTION COSTS

Transaction costs appear when a company makes use of the market to purchase goods and / or services. This includes coordinating the external cost that must be incurred by using the market: ex ante to acquire market information and negotiate a deal costs, and ex post, to prevent fraud and fix it if it occurs. Williamson [49] extensively developed this concept and determined the characteristics of transactions, industries and markets that affect in a fundamental way transaction costs.

The concept of transaction costs can answer the question: what is the role of the market in defining the organizational structure of a company? The answer is that there are organizations and structure to replace the market as allocator of resources, and save transaction costs. Indeed, one can imagine that if there were no costs and information asymmetries market, most of the activities of a company could be performed by subcontracting firms, using the market as a mechanism for pricing outsourcing, making only an assembly of the product or service. However, in reality, the company should incur a series of transaction costs: quotations or bids, contracts, checks, etc. Therefore, in most cases,

One could turn the argument previously expressed, and ask why not use the market rather than the hierarchy. In fact, the current trend towards outsourcing is nothing but a concrete expression of this idea. In addition, theprestigious market is currently as a very efficient mechanism that internalizes a lot of information, production and transaction of products and services; In fact, the price of a product or service contains information not only about their cost of production, but also internalizes trends excess or shortage, price trends for raw materials and labor included in the product, environmental conditions affecting the generation of the product or service, etc. Finally, the market does not require a bureaucracy that explicitly requesting coordinate activities and users, as required by the hierarchy because it automatically induces individuals to take optimal actions, from the point of view of society, pursuing, however, their particular purposes.

The answer to the above question is yes one can use the market in a much more intense than what has been used so far, and for that, there are options of organizational structure involving more or less use.

The first major effect comes from (e-Market) electronic markets, which allow transar produCTOs issued by Internet with access to a much wider range of options, which creates transparency and produces perfect competition, which reduces some of the cost of transaction. The remaining cost, which is related to the implementation of the transaction, may also be reduced using Internet, through automated processes meeting the agreed transactions. However, there are issues such as breach of agreements or quality problems that persist as transaction costs.

For those who want a direct relationship with a supplier, there is also Internet technology to facilitate the transaction. This, in addition to allowing access to sites detailing the offer of many suppliers, offers inteligent agents-implemented entities used in a company or through a site - plying on suppliers and choose products with the best conditions. As in the previous case, once identified one or more suppliers, the implementation of the transaction may also rely on the Internet.

Obviously, the above supports are more viable in standardized products or services. For non-standard products or services, there remains a possibility of using the Internet to identify potential suppliers, negotiate through this medium, with complex exchanges of information such as technical specifications, drawings, budgets, etc. and implement the transaction.

In short, the Internet technology reduces significativously transaction costs and, therefore, encourages the use of the market and tends to jibarizar organizational hierarchies, reducing the vertical integration of enterprises *. This trend is consistent with the idea of focusing a company on what their core competencies and outsource the non-life [21]. This effect has allowed the emergence of many companies besides the electronic-markets exist to facilitate relations through the market.

For example, market information products, particularly content such as TV shows, newsbook columns, comics, electronic courses (e-Learning), opinion pieces, etc. there are companies that act as intermediaries or syndicators between content creators and distributors and consumers, making the package and add functions (bundle) content, and manage relationships between them [48]. This concept also applies to services and physical products, such as unionization of oner site that Amazon.com offers its affiliate program, which enables them to provide hyperlinks from their own websites to Amazon and get a commission for purchases that are verified by this means. There is another idea, very current, which is in line unionization.

Web services, which consist of the provision for Internet -in the form of transparent-any type of service required involving the execution of one or more applications. For example, a customized version of stock market information, several programs that implement the integration function of the supply chain, etc [8]. Companies that provide these services will receive services from multiple sources and provide integrated in a way that is transparent to the user. Net point Microsoft is a technology-oriented in this direction [52].

AGENCY COSTS

On the theory that we present, the owner of a company or its representative is called the principal and subordinate are the agents. That's why this is called agency theory [1, 24, 25]. This economic theory assumes that the assumption of the theory of the firm, in that she behaves as profit maximizer, is too restrictive to analyze the behavior of managers it. The agency theory proposes an alternative vision of a company is a set of related contracts between individuals themselves [25] interests.

In other words, a company is a set of agency contracts, through which a principal (employer) employs agents (employees) to perform some service for him. The assumption that this theory makes behavior-more realistic than the traditional theory of the company- is an agent maximizes their individual utility; he prefers less work and more rewards and do not care about the welfare of principal or other non-pecuniary, such as honor, team spirit, integrity and pride virtues of self-realization.

From the above ideas, we can identify costs that occur within the company and that the traditional theory of the firm no consIdera. First, we have agency costs, which are defined as those occurring as a result of discrepancies between the objectives of the principal and those of agents. For example, consider a business owner distribution of any product, hire vendors to field sales. Sales will increase as the person takes more effort; but every effort unit sales increases in a decreasing amount, or have diminishing marginal returns. The question is what is the optimal contract regarding wages? Suppose, first, a fixed salary. The assumption of agency theory implies that the seller would avoid work hard and sell the amount that would allow him a reasonable effort. An alternative is to give the person a percentage commission on sales. Then the seller maximizes its profits by choosing the level of effort in which its marginal cost (extra effort) equals its marginal revenue. Even with this incentive, the sales may be less than what awaits the owner.

There are other possible solutions to the problem posed agency. For example, the owner can design a contract in which only one payment is made when thesales exceed a certain level, determined such that, when the proper amount of work is applied, such a level is reached; this would ensure that the person would be motivated to apply the right effort, receiving, therefore just reward. However, this is a simplistic scheme as there are other factors affecting sales and have not been considered, such as competitor activity, economic conditions, etc. Alternatively, the seller may return a fixed amount to the owner and keep the rest, if any remaining income. In this case, the risk of the sale is assumed by the seller generally will be less willing to assume the owner. Although the latter two schemes are in the right direction, as to solve the problem of agency cost, they are hardly acceptable by the seller. Finally, the owner can hire someone to monitor the seller all the time (and may need another to monitor the monitoreador and so on). In this case, the principal must balance monitoring costs with increased revenues due to monitoring. In addition, the seller must account sales frequently the owner and document all sales activities, consuming time and effort that could be spent for sale. This loss of time could be avoided if there were no trend behavior minimal effort. Therefore, this is another type of agency cost, called cost of alignment.

The owner can hire someone to monitor the seller all the time (and may need another to monitor the monitoreador and so on). In this case, the principal must balance monitoring costs with increased revenues due to monitoring. In addition, the seller must account sales frequently the owner and document all sales activities, consuming time and effort that could be spent for sale. This loss of time could be avoided if there were no trend behavior minimal effort. Therefore, this is another type of agency cost, called cost of alignment.

The principal must balance monitoring costs with increased revenues due to monitoring. In addition, the seller must account sales frequently the owner and document all sales activities, consuming time and effort that could be spent for sale. This loss of time could be avoided if there were no trend behavior minimal effort. Therefore, this is another type of agency cost, called cost of alignment.

But, despite the monitoring and alignment, the main guilty anyway, in a partial loss of welfare, which we call residual loss, which is the difference between oner expectations and what one actually get.

In short, agency costs are the sum of the costs of monitoring, alignment and residual loss.

Agency costs are further complicated by the separation of ownership and management in the companys, following which it can act according to their interests, at the expense of -shareholders owners, for example. There are also complications due to labor disputes, criminal behavior of employees and conflict of interest between the different managers; for example, between sales and production.

The question is, then, how can a company survive with so many problems. First, direct monitoring of activities is a response. In addition, contracts may be used máso less efficient to direct the behavior of agents; for example, the remuneration of agents may be linked to the result (commissions and rewards for productivity). Also competition, foreign markets and the risk of being absorbed by another company can push managers to favor principal objectives which are above the company-own. On the other hand, external institutions, such as banks, auditing firms and insurance companies, can help reduce agency costs through their own monitoring. Finally, the corporate culture and human nature which possibly is not as opposed to the goals of the company as it implies the theory of agency- may also be directed to reduce agency costs.

However, despite recent factors listed, that can mitigate agency costs, they exist and should be considered when choosing coordination structures internal, as discussed in the above.

But the analysis is not complete if we do not consider, too, how the agency costs and other costs are affected by decentralizing decision rights in a company, which is the design variableIt can be handled within this framework. Of course, if all decision rights are placed at the top of the organizational pyramid -in mainly in theory, at least, agency costs are canceled; and as decentralize these rights, such costs go up. But the centralization of decision rights causes other costs, which have to do with the information necessary for decision-making. First, there are costs associated with transmitting information from where it is generated to higher levels, including communication, miscommunication, opportunity costs due to the delay in communication, etc. This leads to suboptimal decisions by the principal, which is another cost item. Thus, we have another item cost information call costs in decisions, compounds such actual processing costs and opportunity costs due to misinformation. It is clear that if we go down decision rights within the organizational hierarchy, these costs will diminish, because while lower the level of decision, more information is available, it contains fewer errors and more timely.

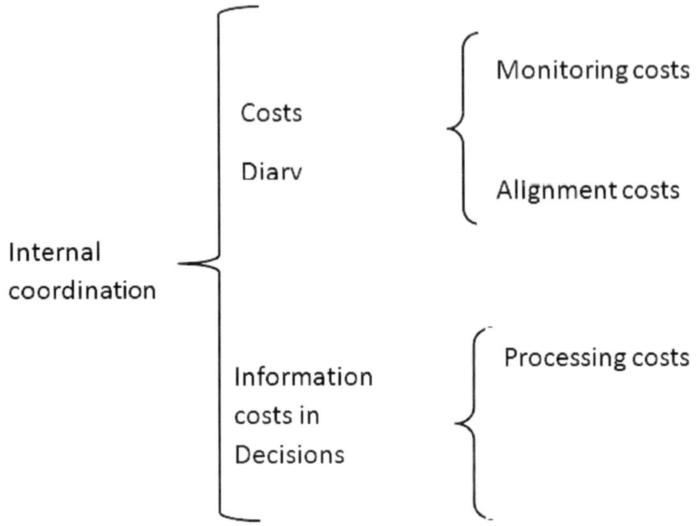

Figure 4.7: Internal coordination costs

Therefore, we are again faced with the need to reach a balance between the costs are summarized in Figure 4.7. This requires locating decision rights where the sum of these costs is minimal.

This indicates that the dilemma centralization and decentralization is false and there is usually a compromise, which is different, depending on the case. For example, in a money table where decisions have to be fast and high opportunity cost, the decision is located at a low level. But to reduce agency costs, remuneration based on results for the agents it sets. At the other end, investment planning in a company is managed centrally, as only the top level may have comparative information between different investment opportunities associated with different agents, and meet budgetary constraints to choose a portfolio optimal investment under budgetary constraints;

A real interesting case illustrating the descentralización of decision rights is the Hewlet Packard, where there are multiple divisions that buy their inputs independently. However, to reduce agency costs inherent in this decentralization-such as not using economies of scale in purchasing and the need to monitor for control- a central database where are registered all suppliers was implemented, corporate agreements supplies and all movements that records each with each of the divisions. Thus, each division decentralized purchasing, but using corporate conventions guaranteeing the overall interest of the company, reducing agency costs.

The current trend is towards descentralition of decision rights, since it has been concluded that a number of costs -such as controls and monitoring reconciliaciones- and opportunity in making decisions -such as approvals and explicit instructions to perform activities that are work- They do not add value and are avoidable, without increasing residual loss. This is achieved by giving more power (empowering) agents - instructions, controls and conciliaciones- but providing the right incentives and ex post controls that prevent residual loss. An example of this type of concept is the Saturn plant, General Motors, where the automobile assembly is done by groups of workers who assemble an important part thereof, without direct supervision, defining their own methods, division of work and tools, and they have veto power over the people within the group, responding only collectively certain goals of quantity and quality [5]. Another example is the already presented Hallmark, where the development of new products has been decentralized to a working group that has all the powers to decide on the design and implementation, leaving the higher levels only monitoring result obtained with the products, which is known, practically, in line with the system mentioned.

Information Technologies enable decentralization, allowing in some cases, also get the benefits ofcentralization. That is the case as presented by Hewlett Packard, where purchasing is decentralized and negotiations and information, centralized, which leads simultaneously reducing opportunity costs in making better decisions -associated and abastecimiento- residual loss, obtaining good prices and conditions at the corporate level. Only monitoring costs due to centralized database to be updated by all participants in the process increases.

Also, Internet technology can help decentralize services and decisions, without the risk of residual loss. For example, a field sales representative supported on a notebook connected to the Internet canhave all the information about the products, their availability and business rules to be applied in a transaction; therefore, may jeopardize sales of their supervisors without intervention, improving cost-avoiding chance- no risk that their decisions are not consistent with the policies and interests of the business service. Also, a branch of a bank can be seen as decentralized, from the point of view of the customer, because with the right systems and communication headquarters can provide all services immediately, almost acting as a selling point. However, it looks after the interests of the bank (principal) having business rules that ensure good decisions, internalized systems using branches.

Agency costs are then affected by the Internet, so that e-Business tend to operate in a decentralized manner with highly automated processes that internalize the polytic and business rules, style and presented cases of Amazon.com and Dell. To these should be added Cisco [46] and Sigma Aldrish, a leader in selling products for laboratories [34], all of which share many characteristics of previous companies.

CHANGE COST

The cost of change is generated in market situations in which customers become captives and have large disincentives to switch supplier of a product or service. The measured disincentive for the cost of change, which includes, for example, the loss of any asset that the customer has purchased as part of the product or service, new acquisitions to do, the cost of training to use the new product or service, and any other costs of adaptation to take advantage of the same.

The classic example of high cost of change is that of software products, particularly those operating systems. Indeed, from mainframe operating systems IBM to the actual Windows Microsoft has been very complicated and expensive for customers to switch to another product. In this case, the cost of change is generated because, in addition to investing in new software, there are considerable costs of renovation or adaptation of all applications running on the current operating system but not in another competitive-and training staff to make it work and operate. These costs can be monumental for a large company that has many computers and applications that run on an operating and a lot of people who use the system. It is exactly this great cost which faces a company that wants to change Windows NT or Windows 2000 to Linux, an option that many are considering today.

Clearly the cost of change introduces rigidities and generates drag on the economy doing less competitive markets.

The cost of change stems from multiple factors listed below [28, 29].

Need for compatibility with existing equipment. For example, components or spare parts required for -computador computer, photocopier, projector- must be compatible with it; other inputs must correspond to existing equipment: ink or toner for printers and leaves for shavers.

Transaction cost to change suppliers. For example, two banks can offer identical accounts, but there is a transaction fee to close an account at a bank and open a competition. It can also be expensive to return leased equipment and lease to a company identical one another.

Cost of learning to use a new product. For example, it is common that computers of different companies are functionally identical, but if a consumer has learned to use the computer with the operating system software mainly, say OS400 for AS-400 IBM- compatible with it, have an important incentive to continue buying equipment from the same company with compatible software.

Uncertainty about the quality of untested brands. Consumers use repetitively solutions that have worked for them and not risk with alternatives that have not been tested.

Coupons and similar mechanisms. Airline passengers register their frequent flyer programs, which reward them for example, free tickets or upgrade to a better class- when traveling repetitively depending on the mileage, which discourages the use of other airlines. Similarly, companies that give away reveal rolls of film rolls that can only be prosecuted for themselves and a supermarket delivery discount coupons that can only be used in it.

Psychological cost of change. While there are no identifiable economic costs to have brand loyalty, may have psychological costs such as habituation or aversion to change.

Contractual costs. In contracts they are signed durable equipment that can bind to a Empresa a provider for long periods of time; for example, a photocopier or industrial printing with a contract that requires buy supplies, spare parts and maintenance from the same supplier, for a good initial discount.

The changes are also costs incurred by suppliers, among which may be mentioned the cost of opening accounts for new customers and uncertainty about the quality of new customers. Whether the company or the customer pays these costs, investment is lost upon termination.

When given in a market described above costs significantly, we examined effects that occur.

The most obvious effect is that the cost of change gives the company poder market on their current customers and creates a potential for monopoly profits.

In particular, Klemperer [28, 29] has shown that individual demand of a company becomes more Inelastica and reduces rivalry with other companies. This leads to a segmentation submarkets, each submarket monopolized by one company.

This leads to a particular form of competition in which efforts focus on comcompetence for market share in the phases of initiation of customers in the product. This implies that prices are lower at this stage as it comes to getting market share that will be valuable in the future because of the monopoly effect. Example of this behavior are banks that give zero maintenance cost or gifts to students in universities to open current accounts; computer equipment offered at a discount to educational institutions to capture students' preferences for future purchases; car companies that accept small gains in economic models to capture customers who can buy more expensive cars after; and discounted insurance policies for new customers.

Competition described can also lead to price wars when a new product is introduced or when a new group of customers entering a market.

Once converted into captive customers, the prices are higher than they would be if there were no switching costs [28, 29].

Another effect of switching costs is that companies have less incentive to diversify, which reduces the variety of products and makes consumers have menos -incurriendo incentives to change in the cost of change- between equivalent products. On the other hand, the companies that sell one version of a product at a disadvantage, as customers, having a high cost of change, prefer a single supplier with a product line; for example the line computer operating system with desktop versions, departmental and corporate network. This favors the existence of companies with product lines.

Finally, the cost of change desencourages the entry of new firms, having them to capture customers reluctant to incur expenses, which further reduces competition.

The information economy naturally creates products with high cost of change. Already it is seen that products such as software markets with captive customers generate high cost of change. Other products such as Zip drive, which create cost incompatibility exchange format- 56K modems which are incompatible with each others; DVD associated with a type of team player, PC and Mac operating system, Novell LANs v / s networks NT and other information products are examples of situations in which captive markets are generated due to the high cost of change.

This leads to a firm that dominates the market is generated and that competition is given by product differentiation, rather than price, after capturing customers have occurred.

NETWORK EXTERNALITY

Externalities in Ddes appear when the utility that a participant gets to participate in a network increases with the number of users of the same. This idea was developed to physical networks, such as telecommunications, which had monopolistic characteristics [32]. However, the most interesting case occurs when several companies compete in a market with these characteristics. This situation can occur in the following ways [26]:

Externalities may be generated by the effect it has on product quality, the number of buyers; for example, the number of holders of a fax or an Internet connection clearly influences the usability of network participants.

There are also indirect effects that generate externalities, such as the one that occurs on buyers of video games, DVDs and the like, in that the total number of buyers determines the availability of content for them. The same applies to the number of computers in a given variety Mac, PC, Sun, etc. in relation to software.

Another form of externality is all durable goods, when the quality of after-sales service depends on the size of the service network, which in turn depends on the number of buyers; for example, in the automotive market, a little known brand is perceived as susceptible to service problems and this slows the growth of its sales.

The key to the existence of externalities is that consumers are on the same network. The size of this network will depend on the type of market. In some cases, as in the automobiles- network it will consist of consumers of a certain brand of a company. At the other end, the network will include all the businesses that sell on the market; for example the market for video recorders.

The feature that determines the size and scope of a network is the fact that products from different companies can be used interchangeably. In communications networks this has to do, for example, the fact that different companies subnets are interconnected and that a user of a subnet can communicate with any other. In hardware, the similar effect is that the software made for a device can be used in other products and durables such as cars, is the set of brands that can share common services.

In markets where there are several networks that compete for the same consumers, they form expectations about the future size of these to decide which choose. This generates externalities demand and, consequently, economies of scale on the demand side. Katz and Shapiro [26] show that, depending on such expectations, only one company will have greater production of zero and, with others, there will be several companies on the market. In other words, if consumers expect a company and its network will be dominant, then they are willing to pay more for the products of the company and this will indeed dominant. This effect is called positive feedback.

Another important question is about the incentives for a company to produce compatible products with others on the market, either through formal or de facto standards. It can be shown [26] that reputable firms or large pre-existing networks will resist compatibility and those with small networks or encourage poor reputation compatibility.

The information economy has the fundamental characteristic the existence of externalities in networks. Now, although there are physical networks like the Internet, virtual networks predominantly determined by the effects described in (a) and (b) of this point. These virtual networks can be for physical products such as hardware / software-for example

Wintel- or information products, such as Internet portals AOL and Yahoo for example.

Positive feedback aforementioned -in the stronger becomes louder explains a lot of market situations where a product has mastered a market obliterating others or leaving them with a very small market share. Classic cases of this type VHS versus Betamax are; Nintendo v / s Atari; Wintel (Intel PC with Windows) v / s Mac; Excel v / s Lotus 1-2-3; NT v / s Novell; and Explorer v / s Navigator.

Unlike traditional economies of scale deals, which have diminishing returns at high volume, complexity of coordination and economies of scale control- demand not dissipate with the size, but increase due to the effect shown in Figure 4.8. A company must move in this curve, in which some reach critical mass and off and others fail.

In virtual networks, in which there are products and the corresponding distribution network more complementary products for example, Windows more Office-, each additional participant who joins positively affects everyone else. Once critical mass is passed, at a cost of collective change that makes very difficult the introduction of competitive products is generated.

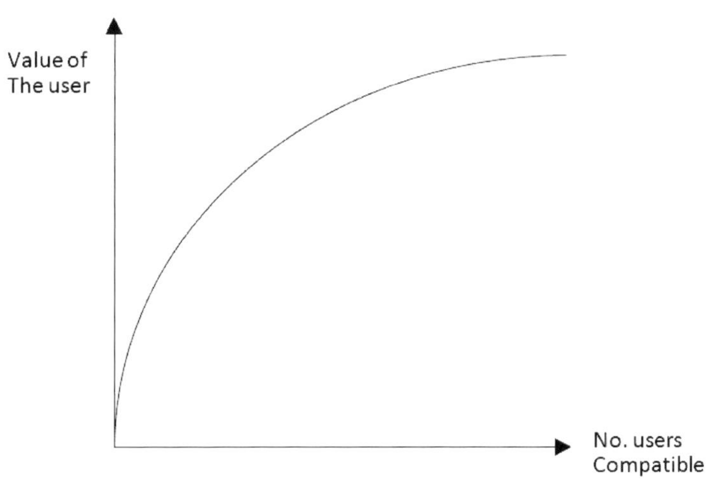

Figure 4.8: Effect of economies of scale demand

When there is standardization of products and anyone can manufacture the positive feedback effect breaks. Products with these features are the PC (as hardware), telephones, PBX and ISP.

Network markets with high and low demand externalities variety of products tend to lean toward one of the competing networks. Having a market this characteristics, standardization strategy may be appropriate for the market launch, given the risk of competing in other conditions.

In network markets occur usually economies of scale of supply and demand, which still favors the emergence of dominant companies in the absence of standardization.

CHAPTER 5

MATHEMATICAL DESIGN OF E-BUSINESS

Ivan Stanimirović

From the foundations elaborated in the previous section, we establish a set of guidelines on how to design business economics information or e-Business. First we propose ideas about the design of the organizational structure of the e-Business and then identify strategic designs to compete in the information economy.

DESIGNING ORGANIZATIONAL STRUCTURE

The organizational structure of an e-Business has well defined characteristics arising from Sections 3.1, 3.2 and 3.3 and the experience of the most successful companies of this type. Then we examine these characteristics, differentiating between companies selling physical products and those who sell information products.

E-Business of physical products

This case includes both traditional companies that have successfully transformed Internet like Cisco, Dell, Lands' End and many others as born with Internet companies; for example, Amazon.com.

The key features of the structure of these companies have to do with the logistical challenge posed by the management of acquisition / production, storage and distribution of physical products in an Internet environment. We examine these features below:

- Well-designed processes and highly automated business. Business processes to which we refer are the chains of interrelated activities that will satisfy the requirements for customer products. Typically they include from customer acquisition activities, through evaluating them and recording orders to satisfy them. But this kind of process is not the only one in business. There are also processes associated with supply chain processes related to the development of new products, processes associated with human and financial resources, and several others.

In 3.2 we saw that the decreasing costs of coordination, particularly those in the information technology support to coordination, encourage high automation supported by computer applications. However, when the Internet is used to serve customers, this automation is a need for processes that implement the back office that meets the requirements of them. Let us consider some emblematic cases to show how it is being treated in practice.

The first case is Amazon.com, whose main process, customer service, shown in simplified form in Figure 4.1. One of the important aspects of this process is that activities 1 and 3 are fully automatic. In fact the "Message Delivery" is generated by a computer application which previously has determined a distribution point from which the pedido- be satisfied by turning on lights in places shelving distribution point where the products requested by a client. In some cases, the Activity 2 runs in a special way to satisfy a customer's order when no stock. Obviously, also runs regularly to replenish the stock of the distribution points. In Activity 4, the employee has only to separate the products, placing them in a box and put it on a conveyor that leads to the despatch area, from where it takes a running courier delivery. To make the process there is support of "Maintenance state", which stores customer databases, and other necessary products in the operation.

There is no doubt that this process could hardly be faster and efficient and is a model that many other companies might copy.

In addition, Amazon.com is exploiting its huge customer base, who frequently visit, to capture orders from other products, whose satisfaction is made by the companies that own products.

One might think that Amazon.com will facilitated the optimization process the fact from scratch, so we then an equally spectacular case of e-Business from the old economy to show that a process one can redesign to carry speed Internet. Dell is the world's leading online computer sales and is also among the top three worldwide sales of desktops and servers.

The remarkable thing about the Dell process is that all computers are made to order and also to automate customer-in similar to Amazon.com- is performed in computational form in Activity 3, scheduling production each individual customer orders, from those released by Activity 1. There is also a high degree of production automation, in Activity 4. Here we have emphasized orders generated online, but Dell also sells over the phone and face to face companies [22]. Furthermore, Activity 2 is performed through the programming providers see Dell production and determine themselves the quantity of components to be delivered and the time they should be delivered. Of course, the management of this relationship is coordinating computer to computer vendor Dell, or automated programming system production client enterprise scheduling system automates production and distribution provider.

Other cases of old economy companies that have redesigned their processes and compete successfully achieved through an e-Business are Sigma-Aldrisch and Cisco, which are very similar to Dell processes. The first sells specialty chemicals for -and also Internet demand traditionally way and is market leader; He received an award for the results of its Web site [34]. Cisco is a company interacting with its customers and suppliers through its website and currently sells 32 million per day, the highest among the sites that sell products directly [50].

The other interesting feature of the e-Business from the old economy just outlined, is that they are tremendously successful from the point of view of economic performance. Not so the dot-coms of the new economy that sell or traded physical products, all of which lose money yet.

The potential for change and improvement in existing business processes of the old economy is immeasurable, as outlined here one are the exception. Most brick and mortar businesses that have a Web site have not yet faced its integration with back office or satisfaction process. A 1999 survey in the US, leading to good use as Internet business, points out that only 25% of companies in the old economy website have made such integration and improved due process [3] What remains for companies in other less developed countries and for those who do not yet have a website?

Now, what can we learn as to redesign and process optimization companies experience outlined above, both the new and the old economy?

The fundamental idea is that the processes have the same structure and the mechanisms that allow optimization are the same type. This can be seen by comparing the situations of Amazon.com and Dell, which are accounted for using, deliberately, the same scheme or pattern. Indeed, both situations share a fully automated service via the Internet, Activity 1 in Figures 4.1 and 4.2, which includes all checks the customer and the ability to provide the product, aided in a database that exists in "Maintenance state", which is updated every time there is a change of status in any of the activities.

But more important than the above, and in line with the idea of attacking the whole process in a coordinated manner, orders are processed in Activity 3, where, by means of an algorithm which is generally complex- is computationally decide the actions to take to meet customer orders, involving, for example, assign productive--bodegas facilities or facilities to meet an order, schedule the sequence of actions in the case of manufacturing packets and determine distribution route in the case of delivery by the company, etc. It is obvious that an algorithm of this type can be derived, in simple cases, based on heuristics with practical basis, or using optimizantes mathematical models in complex situations. Here is the nerve of what is now called the business logic and this determines to a large extent, process performance center.

Decision maker has to computationally decide the actions to take to meet customer orders, involving, for example, assign productive--bodegas facilities or facilities to meet an order, program the sequence of actions in the case of manufacturing packages and determine route distribution in the case of delivery by the company, etc. It is obvious that an algorithm of this type can be derived, in simple cases, based on heuristics with practical basis, or using optimizantes mathematical models in complex situations. Here is the nerve of what is now called the business logic and this determines to a large extent, process performance center.

Program as the sequence of actions in the case of manufacturing packets and determine distribution route in the case of delivery by the company, etc. It is obvious that an algorithm of this type can be derived, in simple cases, based on heuristics with practical basis, or using optimizantes mathematical models in complex situations. Here is the nerve of what is now called the business logic and this determines to a large extent, process performance center.

The third key element of a type process is the existence of a database, represented as "Maintenance state" - which allows each of the activities of the process to have online information required to be executed and updated each time a transaction involving a change of state in the same activities verified.

Finally, there Activity 2, which refers to the supply of raw materials and / or products that the company requires to operate. In both cases, tight integration with the rest of the activities shown, as it aims to buy based on what is needed strictly possibly with a policy just in time- to operate. What is handled here are typical activities considered as part of the supply chain (supply chain management), which currently tends to tight integration with suppliers such as Dell and Cisco do.

The above common elements, presented in very simplified form to facilitate understanding, can arise quite greater detail and to represent a much wider range of situations than those for the two companies exemplified range. This gives rise to what the author calls process patterns that summarize the experience and best practices of many companies in a wide application domain. For example, it has developed a pattern called Macro1, which, in addition to generalize situations like the featured companies also can represent situations in areas as varied as patient care in hospitals, lending to financial institutions and many others (see [7]).

There are cases of e-Business with physical products that are less complex from a logistical point of view than those presented in this point, as selling tickets to shows, ticketing for airlines, electronic markets such as e-Bay, etc. However, however trivial product delivery, there must be an explicit design effort associated process to ensure customer satisfaction.

- Decentralized operation: The costs associated with agency theory -presentados in Point 3.4- push e-Business in the direction of decentralization, since information costs in decisions tend to fall with new technologies, the costs of residual loss, to decrease, by the existence of well-defined rules and automated business and monitoring costs also reduced by use of appropriate technology. Therefore, the operation of the e-Business tends to be highly decentralized. But being internalized business practices in computer systems that automate the operation, as the examples presented in the previous point or support it strongly, there is a centralization in the definition or design of such practices. This is the fundamental factor enabling decentralization,
- Heavy use of the market. As noted in Section 3.3, the Internet technology reduces transaction costs and facilitates the emergence of intermediary agents that facilitate the use of the market, such as electronic markets, sindicalizadores and Web services. Therefore, the e-Business have facilities to outsource everything that is not part of its core business. Cases of this trend are: outsourcing of transport logistics specialists; manufacturing of standardized components of products for example, the components used to build Dell computers by suppliers-; input supply services from other companies such as inputs in a hospital until the level of surgical case made by a supplier or inventory management in supermarkets Wal-Mart Procter & Gamble makes; use electronic marketplaces or auction to find customers or suppliers-for example, Sun's use of e-Bay to top computers or oner Cisco auction site to select suppliers; employee recruitment through specialized websites;

use of back-office services online for example, processing credit policies or through the services of a company located in a foreign country. A notable case which illustrates several of these effects is that sold Nortel 15 parts factories, as integrated circuit boards, and concentrated on integration and value added manufacturing. Manages its suppliers through the exchange E2open site is where one get quotes from potential suppliers. Clearly these methods are implemented through well designed and automated processes, as described in (a).

- Integration with customers and suppliers. This integration is a result of (a), (b) and (c). Indeed, the trend towards outsourcing (c) necessitates a close relationship with suppliers, which is implemented through highly automated (a) processes operating in a decentralized manner (b), with business rules that ensure good practice. As consumers end customers for customers-and that our company is proveedora- integration is mainly due to (a), where highly automated processes that give personalized and intelligent customer service are created, based on a historical record of their behavior.

Now, how they evolve towards a structure with the characteristics described traditional companies that sell physical products from a usually precarious reality in the use of technology.

The usual path followed by companies in the world were as follows:

The first phase is the electronic catalog, that is, through a Web site, disclosed the company's offer, but no transactions, except in a trivial way allowed; for example, a customer places an order for a product, but the process is performed subsequent satisfaction traditionally without automated computer support. This step is relatively simple, but does not provide great benefits to the company, but big problems. Indeed, the cost of capturing orders for entering Internet (transaction cost) and possibly demand increases to be an additional sales channel is reduced, but the slowness of a manual fulfillment process produces a mismatch between customer expectations and service provided,

The second phase is the e-commerce, in which the satisfaction process is redesigned, introducing technology and, as automation explained in (a) -, which will begin to fully realize the benefits associated with Internet sales: low transaction costs, encourages personalized offer more demand, expanding demand for image enhancement and possible extension to other products -for the economies of scale of supply and demand.

The third phase is the E-business, in which complementary electronic commerce processes attack. Among others, the supply chain is attacked, redesigning and introducing practical high automation via Internet; the logistics of the whole business with the same technology is optimized; the processes associated with new product development, management of human, financial and asset resources are attacked; and enters knowledge management, tending to the intelligent enterprise.

The transition between phases can be done gradually, by evolving the same traditional company to the e-Business. An alternative strategy is to set up an e-Business designed from scratch. This is the case of Procter & Gamble which created Reflect.com, adapted to sell cosmetic products (customized) through the Internet and that is completely independent of the array, except for the supply of products. He did it to avoid redesigning a complex enterprise and avoid conflict with their supply chain [48].

An intermediate strategy is to create a separate e-Business, but integrated with traditional customers sharing information and possibly some processes. This is the case of Staples [10], a successful provider of office products created a subsidiary, Staples.com to sell online, which shares facilities and process customer satisfaction with its traditional matrix.

e-Business Information Products

Companies offering pure information products such as Google, Britannica, Napster, etc. have no logistical problems, so the design of its organizational structure is greatly simplified.

The most interesting case is that of companies that offer a combination of physical products and information. Here the trend seems to be convergence of companies offering physical products and information products that offer a single model. Indeed, companies like Amazon.com are evolving from primarily distributors Tailing- -e-business platforms, where what counts and sold is access to the attention of a huge number of potential buyers. In this model, the client is offered a lot of options product on a consolidated basis and mechanisms to search and compare.

Obviously the logistics is discouraged, which is delegated to companies that sell their products through the site in question.

Similarly, some of the companies that sell only content such as Yahoo-are evolving to offer physical products, in much the same way as explained above, where logistics is delegated to end provider site to sell.

In this convergent model, these sales platforms which would be a few-would delegate to suppliers associated with them, logistical problems, which would be very simple structure. On the other hand, providers would adopt schemes as already explained in the previous section, to meet efficiency standards that Internet sales demand.

PRODUCT DESIGN AND PRICING POLICY

Information Products

The high fixed costs and low variable cost of companies marketing products –explained in Section 3.1-, which tends to make them *commodities* It requires a particular approach to product differentiation and pricing policy.

Keys differentiation are customizing products and generation versions, we examine below.

Customization tries to dynamically generate a unique product for each customer of a company. This requires adapting a generic product which has the characteristics of a commodity- to the interests of a client. This is feasible with Internet technologies and business intelligence. Indeed, the Internet technology allows for very detailed information regarding the interests and customer behavior. In particular, one can get demographic information from customers when they register for any type of service or product online or through the information they provide are billed for services or products. One can also document the historical use a registered customer has made a site pages one viewed and / or downloaded, browsing patterns, products purchased, etc. Alternatively, one can set habits unregistered clients through the use of cookie technology.

With this information it is possible to customize directly, by establishing the relevant subset of information that interests a client and offer it proactively; for example, generating a newsletter with relevant information, which is the case of a financial information service which is a commodity- where a company that is interested in the behavior of certain indices or receives information actions so far, only what it is relevant to it, which is the case of information provided by Reuter.

But also to have stories of customer behavior, it is possible to make finer analysis to establish predictive patterns that help customize products or offer certain products to customers that are aligned with their preferences. The technology that allows this is what business intelligence, with variants such as data mining and web mining [20], model-based decision trees, neural networks and others, which identifies the aforementioned patterns; for example, identifying customer segments with distinct characteristics -demographic, preferably for certain products, shop level and others to which they can make targeted offers complementary products, with high probability of acceptance.

Versions generation is closely related to the design of product lines. The fundamental idea is to generate versions for different market segments, each adapted to the needs of a segment. A simple case of this idea is to offer a software version for novices and experts; for example a copy of Adobe Photoshop, called Photoshop Deluxe, sold in conjunction with a digital camera, which can be used by the buyer as inexperienced user. The professional version is for users who acquire a high degree of sophistication and sold separately.

Another example more complex the electronic encyclopedias, where Encarta of Microsoft, with a version of 14 million words and a very low price, managed to have a share of 44.8% in 1995, at the expense of sales of Britannica, which has 44 million words and is much more expensive. Thus, the dilemma was whether to produce Britannica a cheaper version that would allow it to compete on the market under; however, he chose to compete on price. Microsoft countered with a more complete version also low price.

A final example is the version of phone books online companies, which is obviously a commodity. But if a company adds to this a geographical information system to display a map showing the location of a company, it has created a version that has greater value, at least for a market segment.

All the above examples show the fundamental idea that a company must combat the fact that a product becomes a commodity generating versions.

When existing versions, each customer chooses one of them and reveal their preferences in terms of functionality and willingness to pay, because obviously versions have different price. Another traditional example of this -proveniente idea of physical products, but adaptable to information-products are versions of books. These are typically hardback, book back and settlement version of course released in any offset in time.

Variables or attributes versions design can be determined, which are analyzed below [26]:

- Time, ie the time a version, exemplified with books previously released. The idea here is that there are more eager customers receive a product and are willing to pay more for the novelty. Other examples of this type are versions of movies-movies, video, cable, television normally offset in time, and online versions of analysis portfolio of shares and 20 minute delayed in relation to market quotations they use, obviously with different price.
- User interface tailored to the needs of different users; for example, novice users may need minimal functionality and, therefore, an interface very simple. On the contrary, an expert user requires more functionality which leads to a more complex interface. Java applets technology allows a dynamic adaptation of the interface for a particular client, as are programs running on the client machine and can be selected according to the characteristics. Obviously, this technology has the potential to generate an interface and, therefore, a version of an information product for each customer.
- Convenience, which means there is no usage restrictions versions and versions with limitations, of course, with different price. For example, a service that can be used only at certain times, are like some phone plans.
- Quality, which implies that there is a cheaper version of the product having a degraded compared to a more expensive

quality; for example, the image resolution is lower or download speed is slower.

- Functionality, which means a version of a product has less functionality than another; for example, a processor which converts speech to text with a dictionary with fewer words than another.

A very important case versions are magazines on book and electronic versions. The latter can be thought of as more versions functionality of historical material searchable by keyword and material selection of the reader at ease, more opportunity in time, as they are constantly updated; lower graphics quality due to the limitations of the screens; and possibly adapted customer interface. That is, adding and subtracting, the electronic version instead of being free, as it is in most cases should, perhaps, be more expensive if it is well designed. This is the case of the Wall Street Journal.

One of the relevant questions regarding versions is the number of them to design. The simplest thing that can be noted is that very little because it does not exploit the segments, and many increase the cost of product development and confused. Therefore, it is necessary to analyze the market to reach segments identify typical behavior. To do this, the historical behavior of customers is a key element. For example, airlines identify, based on history, segments frequent flyer and, within these, business and tourism, which allows one to design versions of the product passage awards mileage, targeted offers at reduced prices and other mechanisms.

It is important to analyze and identify a product key attributes and manipulate adding value to define products from the top of the line and degrade to generate low. For example, IBM deliberately degraded the performance of Series E laser printer 10 to 5 pages per minute for lower segments oriented by means of artificially inducing release delay periods.

Versions are closely related to the pricing policy. Here idea key -given the low variable cost of the information product is to exploit the propensity to pay for different market segments. One of the notable cases adapted to different segments prices are versions of passages from the airlines, which are quite deliver customized rates to each customer. On the Internet, this can be done online, because when a customer is buying, one can bid for complementary products that are purchasing a custom price; for example books and airfare discount.

In some cases, to get along with customers with high willingness to pay, one have to "spend" to degrade a product and sell it cheaper, it would be cheaper to sell the same product at a discounted price. This is the case of the laser printer IBM previously mentioned.

It has also been found that consumers prefer intermediate versions of the products, that is not the cheapest or the most expensive. This is called "aversion to end" [26]. Therefore, in some cases, it may be advisable to have high and low versions to exploit the above effect without necessarily sold.

A mechanism is also used to define Prices are product packages. These are several products offered together at a certain price, which is obviously less than the sum of the prices of individual products. A classic example is Office package, Word, Excel, PowerPoint and other offers a unique and attractive price compared to purchasing individual products.

The package offers the dispersion decreases the propensity to pay consumers when the propensity to pay for two products a package is not perfect and positively correlated [26]. One can also use the concept of armed packages for the same consumer; for example, an Internet service that generates a tailored CD with a bunch of songs chosen by a user. Another example of packet is given in e-Learning, where one can sell individual courses and programs related courses -secuencias-. These programs can offer a substantial discount compared to taking each course independently. It could also allow a user to define a set of courses of interest and receive a discount for the purchase package.

Obviously this leads, to generalize, to traditional quantity discounts, in this case associated with the variety of products purchased.

Physical products

The interesting case of physical products is that those products are a mixture of traditional and information. For example, a supermarket Internet that offers not only products, but also gives a suggestion service to its users-based on historically information regarding products and also offers discounts based on purchasing activity.

Also a site that offers information about different products and allowing comparisons between them to support a buying decision, including the possibility of suggestions based on previously behavior and implemented after the delivery of products through the same providers.

Clearly, when we add information to physical products they can differentiate themselves with actions of customization and geneeration similar to those explained in the above item versions. In addition, prices may also be adapted dynamically by different segments or even consumers.

Customizing physical products is possible in Internet, due to the large amount of information that can be generated about consumers already explained -of the way for the information product, which can provide useful information to support the purchase and bid online proactively.

For example, Amazon.com makes buying to deliver information about the opinions of others who have bought the same book one are considering and also indicates what other books people bought I acquiredThey ieron the book in question. Another case would be complex from a provider that monitors online inventories and consumption projections of a client company and hands without any intervention is- necessary products at the right time.

The key idea of personalization is to meet both customer through the information we have about it, to be able to anticipate their needs, which involves online marketing for everyone.

Obviously, the price can also be customized. Consider the case of a company that sold by catalog, as Lands' End. This method allowed him only offer the catalog price, more probably, a discount based on the total purchased or other promotion. To transform Internet, prices can be quite dynamic, because, at any time and based on the situation of sale and stock of a particular product can be offered promotions at discounted prices, either on the same site or sending mails offering to customers. One can also hold auctions of products with little movement and generated liquidate.

USING MATHEMATICS TO GENERATE CAPTIVE CUSTOMERS

Strategies to generate captive customers operate in a similar manner for physical and information products, so we make the presentation together.

Shapiro and Varian have identified the cycle that occurs when capturing a customer, which is shown in Figure 4.3 and notes that generates dynamic switching costs and determines its variation over time [26].

The first phase is the selection of brand, corresponding to the event that a customer takes adopts a brand. Examples of this are choosing a DVD to replace a VCR, buy the Linux operating system to replace NT or join a frequent flyer program in an airline. In all these cases begin to generate switching costs will eventually inhibit selection of other brands.

In the sampling phase, the user actively uses the mark corresponding to the product concerned and benefits from the incentives have been offered to induce him to adopt.

This may include providing product samples without cost or free products. For example, a company that delivers a complete version of a software, but with expiration date, or a book club that provides a number of free books against the commitment to belong to the club and to buy a certain number of books over time supplemented with purchased credits for each book that will get new free books. For information products this strategy is very appropriate, because the variable cost of producing a new copy of a product is low. Customers who engage with the product in the previous phase enter the habituation, where they develop a clear preference for the brand in question. This involves completing investment and generate a high cost of change;

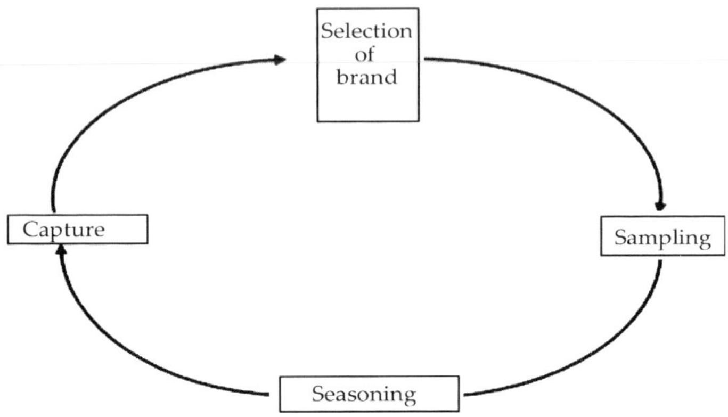

Figure 4.3: Capture cycle

This means that one can have a cost rate high enough to be captive of the brand and with very few options for change. This brings us back to the brand selection phase where, despite switching costs and due to the emergence of attractive alternatives or degradation of the current product, is actively considering other options brand.

The key acquisition cycle is that companies must understand and plan how oner customers will move in it. For example, let us say investment determine discounts or free products demonstration- to be performed for customers to choose the brand and move quickly to Habituation. This can be calculated by assessing the value of a portfolio of captive customers, which has to do with the net profit generated by a client with this feature during their life cycle. That is, the fundamental idea is to invest to capture. This investment is related to the factors causing the cost of change explained in Section 3.5.

The cost of change and customer acquisition can be analyzed from the point of view of the user and the provider.

From the point of view of the company, person be the captive is obviously not desirable, especially when it is tied to inferior products. Therefore, it is desirable to avoid capture, which can be achieved by opting for standardized or products, but these do not exist, open products. These are those who have public specifications and for which the technology involved in its manufacture does not own; for example, Linux or software products built to the CORBA standard.

Complementary to the previous approach is to keep costs under control change, by always having an open migration path to other brands; for example, choose a package manager relational databases from a particular supplier, but use languages SQL and C-say standard programming and not be tied to a vendor with stored procedures programmed on a proprietary language him.

When one can not avoid capture, try the best party to it one get. This can be achieved by performing a very good negotiation prior to capture. This should include aspects such as discounts, extended warranty, support for change from the previous technology and benefits throughout the cycle for example, free updates in the case of software.

Now, from the point of view of suppliers, customer acquisition can be a very good business and should therefore be attempted. The fundamental idea and big business is to create products or even proprietary architectures protected by patents- to be adopted massively and become de facto standards, such as Windows in the PC world.

Some possible strategies for achieving this are discussed next.

Investment in captive customers. First, we must recognize that suppliers and customers negotiate to reach an agreement on the benefits and conditions that obtain the latter to commit to a brand, taking first into account the returns obtained when customers are captive. This can be thought of as a zero-sum game not because both can benefit from the agreement. In fact the buyer gets discounts to offset the cost of change and seller, which differs income will be higher than if customers were not captives for subsequent phases of the cycle.

Within the framework of previous negotiations, the first thing to consider is how much the provider to invest in building a portfolio of captive customers and also the maintenance cost. We said that in determining this value should look at the whole cycle of capture, explained above, which involves considering all income expected from a captive customer, from upgrades, maintenance, complementary products, supplies, etc. In the background the idea is to consider customers as an asset and determine how much one can invest to get a new customer. This idea implies that the returns to be obtained from captive customers, possibly at higher prices than they would in perfectly competition are not excessive, and they will pay the initial investment. This is true provided that had jurisdiction in the initial phase cycle arrest [26]. A mistake that we must try to avoid is not overinvestment in capturing customers because they get a high market share with captive customers docs not cnsure a high cost of change and, therefore, the maintenance. For example, there is a risk that a substitute product sufficiently compatible with ours as to reduce the cost of change appears. A case where this has been tried, unsuccessfully still is with Linux, Gnome and StarOffice combination -all free or very low cost products which, in combination, provide an interface similar to Windows and Microsoft Office-like functionality, along with compatibility with files of the latter. The successful case is that of Explorer, which displaced Navigator,

The problem of attracting new customers and turn them into captives is complicated when they are customers of other brands and have high cost of change. In such a case, it is essential to calculate precisely such costs and subsidize form, provided that the overall economic analysis justify such investment based on the benefits to be gained once customers become captives. Another way to get captive customers is trying to sell to influential customers, who are perceived as leaders or control standards. Finally, one can try a multi-player strategy, as to sell to a customer who drags others. For example, an airline sells tickets on sale -in very conveniently form a company for business trips of its employees,

Maintenance of captive customers. Once captured customers, the challenge is to keep them, which can be achieved in several ways. First, the design of the products may have proprietary features that make it difficult migration to other products; for example, proprietary formats, programming languages, as the owners of stored procedures administrator's database systems and parts Relacionales- Data unalternative some machines such as the ampules of retroproyectoras files can also be associated information services add value to physical products, which generate costs change. This is the case of a supplier of standard inputs for which hospitals, obviously no cost exchange itself-giving a value added service maintenance of inventories of hospitals, taking care of the periodic replacement and providing them to the level of surgical case, which, of course, generates a unique product with high cost of change [22]. Another way to keep captive customers is to introduce loyalty programs and cumulative discount, from their history. This strategy makes any traditional market, at no charge exchange, in a market with captive customers. Book clubs with credits earned for each book purchased, valid for purchase other books free of charge, voucher programs in supermarkets and gains points for product purchases with credit cards, are examples of such programs. Finally, one can control the cycle capture a customer through early and updates renewable contracts, for example. In the same vein, one can establish mutual agreements -game convenience zero-sum between suppliers and customers, where they can emphasize stability, quality and security of supply of products.

DEVELOPMENT OF CLIENT NETWORKS

Clearly, from the point of view of the provider, taking over a market by making dominant its customer network, is strategy attractive. At least, much more attractive than one of the networks that disappears or is reduced to a minimum, according to what usually happens in networks with high externality.

The other possibility, in addition to the two end up Sketchingdas, in a market with high economies of scale in -typical demand for the products of information- economy it is to agree a standardization of products with other suppliers, either formally or de facto. This has happened in the market for DVD players, where the format is standard, and the market for UNIX operating systems, although a standard that is not fully respected by manufacturers-sharing software applications developed for them.

Creation of a dominant network

In order to build a dominant network with a certain product, creating a market with high positive feedback, it is possible to use two estrategias: evolutionary or revolutionary. These strategies can be conceptualized in the way shown in Figure 4.4. [26].

In evolutionary strategy compatible product from the competition is designed with a performance superior to hers, but not extraordinary because of the limitations imposed compatibility. This is to minimize the cost of changing customer and, therefore, facilitate migration. This was the strategy followed by US Robotics with Palm product, although using a different operating system dominant, which is Windows, has the ability to exchange files with him, which has allowed this computer calendar or clones of the same master roundly market mobile computers.

Migration from other products can be achieved in various ways; for example, a -founded very creative design in a very good engineering-, which develops functionalities or not present efficiency in existing products. So Palm design emphasizes simplicity of use and small size and high performance software.

Another way to attract customers is to design in systemic terms, thinking about the products provided by our company and others. This has been exploited by Palm and his followers leaving open the possibility that many other developing software products and hardware, such as programs that offer additional functionality -planillas computing, gaming, financial, etc. I components are coupled to through a universal interface, such as cell phone, MP3 player and others. Finally, a company can attract customers by creating bridging technologies such as converters and emulators. At the Palm, as already mentioned, one can convert files of certain functions of it the equivalent of Windows and vice versa. A similar case is allowed to read Excel spreadsheets Lotus 1-2-3 and Word that allowed read text files Word Perfect. Another is Linux's Gnome-a software that provides for a Windows- like interface and StarOffice Microsoft Office--a clone of that together, provide similar functionality and less expenditure of resources that Microsoft products equivalent and can read files of these products, facilitating the conversion. All these products are virtually free.

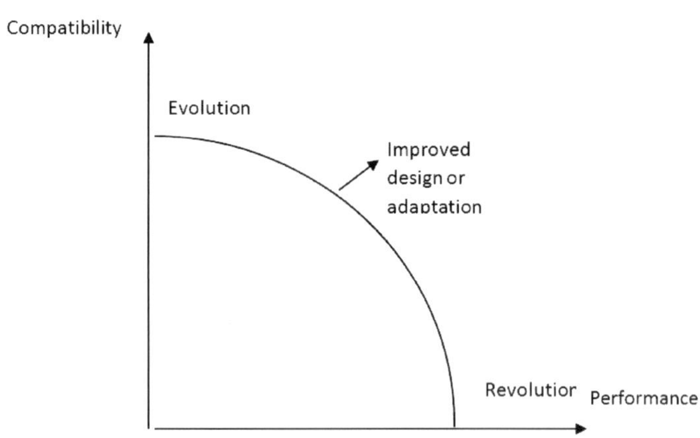

Figure 4.4: Compatibility v / s performance

The alternative to the evolutionary approach is revolutionary. In this extraordinary performance it is pursued, but incompatible with current products, as shown in Figure 4.4. Increased performance should be significant enough to induce consumers to assume the current high cost of change and those entering the market, prefer them clearly. An interesting case of using this strategy the product Nintendo v / s Atari, where the first developed a superior technology that made the second customers and new consumers will prefer.

This strategy is obviously very risky, because the investment to get a superior product and marketing to place on the market is large and there is no assurance that the product is preferred in the market. Famous cases of products that failed in this attempt are OS / 2 IBM network operating systems and MDs.

Once a product with appropriate features for creating oner network externalities customer is accepted in the market and acquire critical mass, economies of scale demand and positive feedback will take to dominate the market. Once in this position, there are several ways in which a company can capitalize on their installed base of customers.

An important opportunity to benefit from a network of clients is selling complementary products, which now involves not only sales but also in the future, such as maintenances, upgrades and extensions. This is the strategy followed by Microsoft, once appropriated market desktop operating systems with Windows, introducing Excel, Word, Powerpoint and eventually a product package all previous: Office. In addition to selling products such multiple updates, Microsoft has culminated in Windows 2000, which, in various versions, try not only cover the desktop market, but also the local and corporate networks.

Another example of selling complementary product to monetize an installed customer base is that banks have followed with credit cards. The network is created with a product that has a relatively low margin, but once built, one are offered to customers other high-margin products such as credit cards and associated products other banks.

Another strategy to value a customer network is selling access to them so that other suppliers sell them their products. For example, Amazon. com that pays its millions of customers selling toys and electronics for other providers, paying them a commission and running with the logistics associated with delivery. A similar case is what Yahoo is doing, selling products for others and is what made AOL, selling access to its customers Amazon.com. In all these cases is selling knowledge that companies have about their customers and, obviously, to the extent that detailed knowledge of their behavior is like one can get through Internet-, one have this possibility will be higher .

One can also use a strategy tiered priced to get more out of an installed customer base. The idea is to discriminate according to the knowledge we have of customers. Thus, historical customers with high cost of change and subject to high network externalities, will be willing to pay regular, obviously subject to the promises and agreements that exist with these prices. On the other hand, customers of competition have to charge them a price that discounts the cost of change, otherwise the can not capture, and new customers have to offer introductory prices.

This strategy of differentiated prices should not discourage old customers, so one have to keep track of their response to pricing policy. Furthermore, complementarily, can be used versions of the product; for example, versions with higher added value for old customers who pay full price.

Collaboration

Not always the strategy of trying to dominate a market fully taking advantage of economies of scale and network feedback is positively the most appropriate. The important thing is to maximize the total benefit obtained, whose balance is shown in Figure 4.5 [26]. It follows that the total profit for a company is:

Total profit = total value added of the x market participation

a company industry business

Therefore, there is a balance between a proprietary strategy, with a superior product protected by patents that tries to dominate the market- and an open, in which standards are agreed with other companies and competing on equal terms.

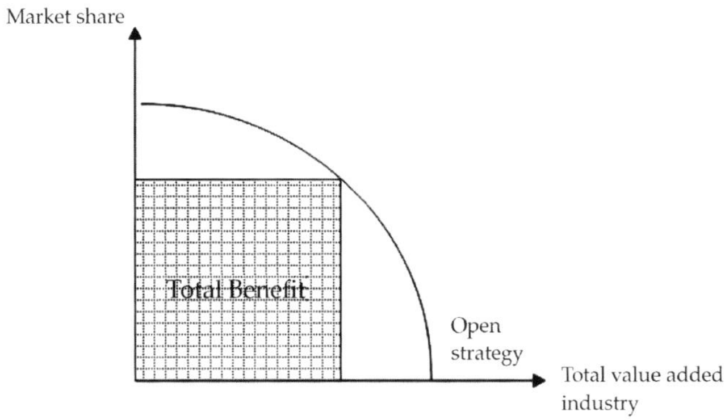

Figure 5.5: Benefit of a marketing strategy

Trying a proprietary control has the advantage of capturing customers and making them profitable in the manner already indicated, but can inhibit the product off for lack of openness. This was what happened to Betamax, which when trying to control a proprietary technology, discouraged the cooperation of other companies and created the conditions for the emergence of a successful alternative technology, VHS. The key is that the total value added of the industry depends on the network effect and this, in turn, the degree of opening, as shown in Figure 4.5. Therefore, when no company can impose standards and this prevents the network dynamics is generated, it is necessary that an alliance between providers formed to agree on standards and develop the market. This is the case of Java, though some criticize that the product is not fully open because Sun controls the changes; digital HDTV; and ATM networks occupying banks. This is particularly relevant when network externalities are extremely strong and no single company can develop a critical mass of customers.

The alliance involves cooperation between providers to create a single network, generating a combination of cooperation and competition (coopetition). This aims to create national or international standards that enable compatibility or interoperation between products from different vendors, creating value for the user, as there is a larger network. Manufacturers and consumers favored their interests for the elimination of technological risk, reducing costs and improving early expectations seconds. Obviously, switching costs -due to interoperatividad- favoring consumers are also deleted.

Therefore, competition between providers is-not market share dominance with owner- technology, favoring consumers, again, with lower prices or better value-added services; for example, support.

This strategy allows differentiation with proprietary extensions, which can be obtained additional benefits. For example, in the market for database administrators - which share a relational database that is standard SQL and possibly other languages also could be supported as C- a single network in which applications could be exchanged between different brands software. However, manufacturers have developed complementary products such as proprietary languages, development aid and others who produce incompatibility and generate additional income to segment the network.

Important examples of collaboration between providers to agree on a standard are the PC market, which has become a commodity-, that of VHS, the discs 31/2 "and the aforementioned digital television high resolution and ATM terminals. Recent examples are the DVD and XML.

CHAPTER 6

MATHEMATICAL MODELS IN CAPITAL MANAGEMENT

Ivan Stanimirović

The working capital management refers to the management of all current accounts of the company including all assets and liabilities, this is an essential point for management and financial regime.

Management of company resources are essential for progress, because it is this which we largely measured the level of solvency and ensures a reasonable safety margin expectations for managers and administrators.

The primary objective of capital management job is to manage each of the assets and liabilities of the company such that an acceptable level of this is maintained.

The main current assets to which they should pay attention are cash, marketable securities and investments, accounts receivable and inventory, as these are the ones who can keep a recommendable and efficient liquidity without retaining a high number of stocks each, while liabilities are most relevant accounts payable, financial liabilities and accrued liabilities because these sources of short-term financing.

When a company has cash uncertain inputs must maintain a level of sufficient current assets to cover its current liabilities.

Working capital can be defined as "the difference that occurs between assets and current liabilities of the company". One can say that a company has a net working capital as current assets exceed its liabilities in the short term, this implies that if an organizational entity wants to start a business or production operation must handle a minimum working capital will depend on the activity of each.

The pillars that management of working capital is based are based on the extent one can do a good handle on the level of liquidity, because the more wider the margin between current assets held by the organization and its liabilities increased circulating is the ability to cover short-term obligations, however, a major drawback occurs because when there is a different degree of liquidity associated with each resource and each obligation, when unable to convert current assets more liquid money, the following assets have to replace them because the more of these will have the more likely to take and convert any of them to meet their commitments.

Modeling of the working capital

The origin and the need for working capital is based on the environment of the cash flows of the company that may be predictable, also based on the knowledge of the maturity of obligations to third parties and credit terms with everyone, but in fact what is essential and complicated is the prediction of future entries to cash and assets such as accounts receivable and inventories are items that in the short term are difficult convertibility into cash, this shows that among more predictable future cash inflows, the lower the working capital needed by the company.

The primary objective of capital management job is to manage each of the assets and liabilities of the company.

Profitability vs. Risk

It is said that "higher risk, higher the return". This is based on the management of working capital at the point that profitability is calculated by profits after expenses against the risk that is determined by the insolvency which may be the company to pay its obligations.

A concept that takes strength right now is how to obtain and increase profits, and theoretical foundation is known that for an increase in these two essential ways to achieve this, the first is to increase revenue through sales and secondly reducing costs by paying less for raw materials, wages, or services to be provided, this postulate is essential to understand how the relationship between profitability and risk together with an effective management and execution of capital of work.

"The greater the amount of working capital that a business, the less the risk that this is insolvent," that is based on the relationship that occurs between liquidity, working capital and risk is that if They increase the first or second third decreases in equal proportion.

Already considered the above, it is necessary to analyze key points to reflect on proper management of working capital versus maximizing profit and minimizing risk.

Nature of Business: It is necessary to place the company in a context of social and productive development, since development of financial management in each is different treatment.

Capacity assets: Companies are always looking for naturally depend on its fixed assets in greater proportion than in current to generate profits, since the former are those who actually generate operating profits.

Financing costs: Companies obtain resources through current liabilities and long-term funds, where the former are cheaper than the latter.

Consequently management of working capital has variables of great importance that have been discussed above quickly but concise, each are a key management made by managers, directors and managers of financial management point is recurrent then take all necessary measures to establish a financial capital structure where all current liabilities finance effectively and efficiently current assets and determining optimal financing income generation and social welfare measures.

The financial manager can establish a policy for working capital to vary from very conservative to very aggressive.

A conservative policy would put more emphasis on risk minimization (being very liquid) as opposed to maximize yields.

An aggressive policy would emphasize the aspect of yields.

CASH MANAGEMENT

Cash generation is one of the main objectives of business. Most of its activities are intended to result in a direct or indirect way, an adequate flow of money that, among other things, finance the operation, investing to sustain growth of the company, pay, where applicable, liabilities maturity and, in general, to compensate owners satisfactory performance.

In short, business is business only when it generates a relatively sufficient amount of money.

Some ideas and practices that are currently driving some companies to achieve effective management of their cash are as follows:

One CAN NOT MANAGE WHAT IS NOT KNOWN.

The first requirement to manage cash, is to know both how it is generated, and how it applies or destination is given. This information is provided by the cash flow statement. This state reports, on the one hand, what have been the main sources of money from the company (where he got the money) for a certain period of time and moreover, he has done with it (in which he has applied).

The money generated primarily by charges to customers from the sale of inventory or providing services, less payments to suppliers and expenditures to cover the necessary costs to sell, buy and produce goods sold during the period covered by the information is the net cash generated by operations.

The ability to generate cash through operations, to determine, among other things, the potential growth of the company and its financial health.

In addition to this source, as we know, there are others:

Liabilities.

Contributions by shareholders (capital increases).

Sale of investments.

In addition to the operation, the cash can be applied to:

Pay debts.

Pay dividends.

Repay capital to shareholders.

To invest.

Investment in assets (customers, inventory, fixed assets) is “stored” money. Hence there because finance the aphorism: “Before one go to apply for a bank loan looking money one have tied up in the portfolio in stocks or other assets.”

WHO IS RESPONSIBLE FOR MANAGING CASH FLOW?

As it was noted, most of the activities of the company are intended to affect, directly and indirectly, the flow of the company.

Consequently, its administration is a task in which they are involved all people working in the company. What each individual does (or does not do) will affect one way or another company cash.

THE FOUR BASIC PRINCIPLES FOR EFFECTIVE MANAGEMENT

There are four basic principles whose application in practice leads to proper management of cash flow in a particular company. These principles are aimed at achieving a balance between the positive flows (cash inflows) and negative flows (outflows) retail so that the company can consciously influence them to achieve maximum benefit.

The first two principles relate to the inputs of money and the other two outlays of money.

FIRST PRINCIPLE: Whenever possible should increase cash inflows. Example:

Increase sales volume.

Increase the sales price.

Improving sales mix (boosting higher contribution margin).

Remove discounts.

SECOND PRINCIPLE: Whenever possible should accelerate cash inflows. Example:

Increasing cash sales

Ask advances to customers

Reduce credit terms.

THIRD PRINCIPLE: Whenever possible should reduce cash outflows. Example:

Negotiate better conditions (lower prices) with suppliers.

Reduce waste production and other activities of the company.

Doing things right the first time (Lower costs of not having quality).

PRINCIPLE FOUR: Whenever possible should take the cash outflows. Example:

Negotiate with suppliers the greatest possible time.

Acquiring the inventories and other assets closer to when they will need time.

It should be noted that the application of a principle can contradict another, for example: If sold only in cash (canceling credit sales) is achieved accelerate inflows of money, but the risk of a decrease in sales volume runs. As one can see, there is a conflict between the application of the second principle and the first.

In these and other similar cases, one must assess not only the direct effect to the application of a principle, but also the additional consequences that can affect cash flow.

OPERATION IN PRACTICE OF EFFECTIVE MANAGEMENT OF CASH

Some companies have gained important achievements in its cash management, applying these four principles, following a more or less similar procedure as described below.

To achieve this goal, it is necessary to first explain to the participants in the meeting as cash generated and applied in a company. Must be achieved quickly to detect that cash flow is everyone's responsibility, as well as other relevant aspects of cash management.

Next to this stage induction, it is recommended that each participant individually and structured, reflect how each of the main activities carried out in their work affect and are affected by the cash flow of the company.

This reflection is intended that each individual "land" his own reality, direct or indirect, in the cash flow of the company participation.

Once this is achieved, the participants into small groups (three or four people in each) is divided, to identify symptoms team, define problems and offer real solutions to work situations. After each subgroup did his homework, the entire group is collected and shared and discussed the problems involved.

Both individual reflection and seeking and discussion groups, should be referred to the application of the four principles of cash management. This implies that responsa each of the participants, what one can do (or not do) is:

Increase cash inflows.

Accelerate cash inflows.

Reduce cash outflows.

Delaying cash outflows.

The final product of the meeting, a list of detected problems and possible solutions proposed is generated.

CASH CONVERSION CYCLE

Is the time between the payment of accounts payable and cash receipt accounts receivable, ie they are the days between the date on which the company must start paying suppliers and the date starts receive cash from customers.

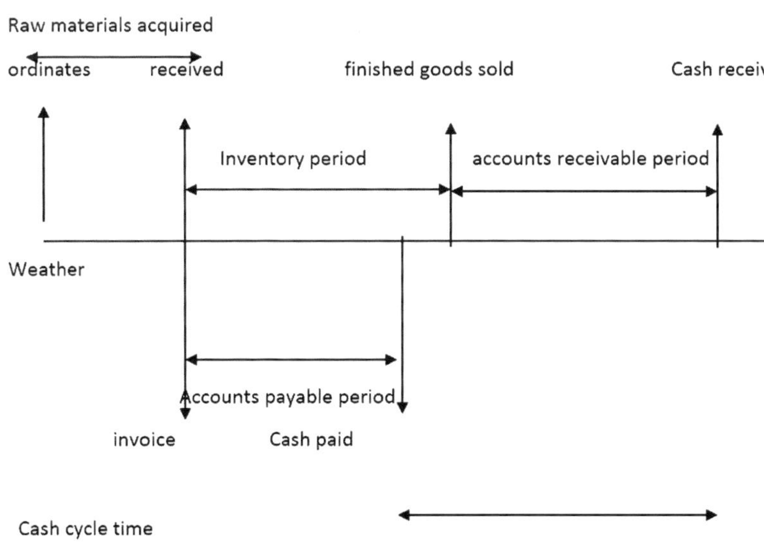

Cash conversion cycle	=	Inventory conversion period	+	Collection period of receivables	-	Deferment period accounts payable

The inventory conversion period is the average time between inventories are bought and sold goods.

$$\text{inventory conversion period} = \frac{Inventory}{Cost / 360}$$

The collection period of receivables, or days sales outstanding (DVP) is the average number of days required to collect accounts receivable number.

$$\text{The collection period of receivables} = \frac{Receivable\ accounts}{sales / 360}$$

The period of deferral accounts payable is the average time between the purchase of materials and workmanship that goes into inventory and paying cash for these materials and workmanship.

MODELS OF DEMAND FOR CASH TRANSACTIONS.

How much cash one need an organization to conduct transactions? A simple and useful approach to answer this question about cash management as an inventory control problem. With this approach, the company manages its cash inventory based on the cost of holding money (instead of marketable securities) and the cost of converting cash securities. The best policy minimizes the sum of these costs.

BAUMOL MODEL

The Baumol model assumes that cash management:

The company can predict with certainty its future cash needs,

Cash disbursements are uniformly distributed within the period,

The interest rate (the opportunity cost of holding cash) is fixed and

The company pays a fixed transaction cost each time converted into cash.

The Baumol model has a pattern of cash balances resembling saw teeth.

The company sells C weights of marketable securities and deposits the funds in a checking account. The cash balance is continuously reduced to zero, as the cash is spent.

Then the company sells C weights of marketable securities and deposits the funds in the checking account, and the pattern is repeated. Over time, the (average) average cash balance is . The time value (opportunity cost) of funds is the interest rate, i, multiplied by the average balance. Also, if each deposit is C pesos and the company needs to deposit a total of T pesos in oner account during the year, the total number of deposits is equal to T / C. The annual cost of the cost per transaction deposit multiplied by the number of deposits, ie b T / C.

The annual cost of meeting the demand for transactions is the transaction cost plus the opportunity cost (time value):

$$\mathrm{Cost} = \mathrm{b}\left(\frac{\mathrm{T}}{\mathrm{C}}\right) + \mathrm{i}\left(\frac{\mathrm{C}}{2}\right)$$

Where:

T = Annual volume of transactions in pesos (uniform over time)

b = Fixed cost per transaction,

i = annual interest rate,

C = Size of each deposit.

The variety of choice is C, the size of the deposit. Increases C increases the average balance of cash and the opportunity cost, but the number of deposits, bringing the annual cost decrease transaction. The optimal size of deposit, C *, gives rise to a minimum total cost, cost *.

The optimal size of deposit is:

$$C^* = \sqrt{\frac{2bT}{i}}$$

CASH MANAGEMENT MODEL MILLER-ORR

The model cash management Miller-Orr is more realistic than the Baumol model, as it allows daily cash flows vary as a function of probability. For example, suppose that we can assume that the expected net cash flow (inputs and outputs combined), follows a normal distribution with a measurement of zero and a standard deviation of 50000 per day. This realistic uncertainty about future cash flows is incorporated.

The Miller-Orr model employs two control limits and a return. The lower control limit, LCI is determined outside the model. The upper limit, LCS, this 3Z above the lower control limit. The model uses a return point Z that is above the LCI. The turning point is the target level at which the company returns whenever the cash balance reaches a limit control.

When the cash balance drops to LCI, the company sells the amount Z of marketable securities and place the money in the checking account. This makes the cash balance to rise to the point of return. Whenever the balance up to the upper limit, bought 2Z of marketable securities with money from the checking account, and this makes the balance falls below the point of return. This procedure produces an average cash balance of LCI + (4/3) Z.

Although LCI is set outside the model, LCS and the return point depend on the LCI and Z. The return point is simply LCI + Z, and the LCS equals LCI + 3Z. Z variable depends on the cost per transaction, b, the interest rate per period, i and the standard deviation of cash flows, σ. Interest rate and standard deviation of cash flows should be defined for the same unit of time, which is usually a day. Z is then:

$$Z = \left(\frac{3c\sigma^2}{4i} \right)^{1/3}$$

As noted above, the point of return, PR, the upper control limit, UCL and the average cash balance are:

PR = LCI + Z

LCS = LCI + 3Z

Cash balance average = LCI + (4/3) Z

A higher cost per transaction, or a greater variation in cash flow, produces a larger Z and a greater difference between LCI and LCS. Also, a higher rate of interest produces a smaller Z and a narrower range.

The Miller-Orr model is a good example of management by exception. Managers allow cash balance fluctuate within certain limits. Managers act only when these limits are violated, and that action is planned in advance. The model is useful for managing cash balances for transactions.

COMMERCIAL DISCOUNTS

It is essential that the company has enough cash to accept trade discounts. The payment program for purchases is called term sale. A billing procedure commonly found, or term trade, is one that refers to a discount on a bill paid within a smaller number of days that that which is required full payment to date expiration.

Suppose one are offered a 2% discount on a bill paid within 10 days, requiring full payment within 30 days. Since the net amount due within 30 days, stop taking the discount means paying an additional 2% for using the money 20 days. The following equation cost, on an annual basis, of not taking the discount is calculated:

$$cost = \frac{discount\,\%}{(100 - discount)} \times \frac{360}{(Expiry\ date - period\ of\ discount)}$$

$$\text{cost} = \frac{2}{(100 - 2)} \text{ x } \frac{360}{(30 - 10)}$$

$$Cost = 0.3686 = 36.86\%\ annualy$$

Then we determine the real rate of interest:

$$real\ rate\ of\ interest = \left(1 + \frac{\text{cost}}{\text{t}}\right)^{t} - 1$$

where t=(Expiry date-period of discount)

Therefore, actual rate is equal to 0.44 and the real rate is 44% (real annual rate).

This represents an annual real rate of interest. The capital cost of many companies is substantially lower than that obtained with the formula, and must therefore borrow funds, if it necessary to take off when the implied interest rates on cash discounts are of magnitude greater.

INVENTORY CONTROL

Inventories play a crucial role in the process of purchasing-production-marketing company. Some are a physical need for the company. For example, assembly lines, must be partially built cars. Pipelines must contain oil in transit. Other inventories are reserves "cushion" needed at various points of the procurement process-production-marketing. If foods are not in the grocery store when customers want, the store can not sell them. Also, if the necessary parts or are slow to come out, the assembly line could stop. Some goods such as grain or coal (coal, ore, fuel), embark on such amounts may be required one year or longer to clear a shipment.

Manufacturing companies generally hold three types of inventories:

Raw Materials

Production process and

Finished product.

The size of inventories of raw materials of a company depends on factors such as the expected level of production, production change with the season and the reliability of supplies. The size of the inventory company process mainly depends on the overall length of each production cycle, and the number of different stages that constitute the cycle.

Finally, the size of inventories of finished product of the company depends primarily on the rate of sales (units of product per unit time), the cost of maintaining inventory, ordering cost of stock replacement and the cost of that an article (and such sale is lost or until the client) runs out. In general, larger inventories give the company greater flexibility and operational sales, but also cost more.

MODEL EOQ

A control model simple and useful inventory is the economic order quantity. This model is derived as follows. Suppose inventory units are removed with a speed S (how quickly the product, in the case of inventories of finished goods sold). Suppose there is a fixed cost of replenishment, F, per order, regardless of the number of units ordered, and it costs C maintain inventories unit for a whole period. It should be noted that the model **EOQ** is a constant use of inventory (or steady sales) and instant replenishment of inventories.

Q initially contains inventory units and constant speed is reduced to zero. At that point, the inventory is replenished instantly with other Q units, and the process restarts. Throughout the year, the inventory fluctuates between Q and zero.

The total annual cost has two components, refill costs and maintenance costs.

The cost of restockingis the cost per order, F, multiplied by the number of orders per year. The number of orders per year is the annual expenditure (in units) divided by the size of the order, S / Q, so the annual cost of replenishment is:

$$F\left(\frac{S}{Q}\right)$$

The maintenance cost Annual maintenance is the cost per unit, C, multiplied by the average inventory, it is Q / 2. Thus the annual maintenance cost is:

$$C\left(\frac{Q}{2}\right)$$

The total annual cost is the sum of two components:

$$\text{Cost Total} = \text{Cost of refill} + \text{Cost of maintenance} = F\left(\frac{S}{Q}\right) + C\left(\frac{Q}{2}\right)$$

We can minimize the total cost if we find the order size that balances the two components costs. Order that size is the economic order quantity (CEP), the formula is:

$$\text{CEP} = \sqrt{\frac{2FS}{C}}$$

DISCOUNTS FOR WHOLESALE

Many providers offer a discount for wholesale to encourage larger orders. For example, a provider might offer a price discount on orders of 10,000 or more units. If our current CEP is greater than the carrying amount discount, the discount is obtained without having to do anything.

If the CEP is less than the carrying amount discount, it is necessary to increase the size of order to get the discount. The balance between the costs of higher inventories and lower price of purchases. If the savings exceed the cost of extra inventory is recommended to increase the size of order to obtain discounts. To analyze this decision, we can adjust the total cost function by adding the price discounts:

$$= F \frac{S}{Q} + C \frac{Q}{2} - dS$$

Total cost = cost + cost refill maintenance - price discounts

where d is the discount unit.

The size of optimal order is what gives rise to the lowest total cost: either the result of using equation (1) for the CEP calculated without discounts on the price, or the result of using equation (2), we it gives the cost of ordering more and get discounts.

INVENTORY CONTROL WITH UNCERTAINTY

The CEP model includes assumptions that simplify the calculations: future demand is uncertain, inventories are spent at constant speed and delivery is instant (or, equivalently, the lead time is uncertain). Each factor, of course, has some uncertainty. A company can protect themselves from this uncertainty by keeping safety stock, ie, a "cushion" inventory.

The company spends inventories to the point refill. At the point of replenishment, order oner CEP, but there is a time of anticipation until the order arrives. During that time ahead, the company stocks may run out if the demand is great. There is a situation out of stock when the company can not immediately make a sale due to lack of inventory. This situation can harm the relationship with customers and / or cause lost sales. To guard against this, the point refill includes expected demand during lead time plus safety stock.

Refills point = expected demand in the lead time + safety stock

If there is uncertainty, the cost of annual inventory has three components:

Total cost = cost of maintenance cost refill + + cost of stock

The expected cost of the stocks are exhausted is the probability of exhaustion times its cost. A larger security presence reduces the expected cost of exhaustion but raises the point refill. Thus, a greater security presence increases the cost of maintaining inventory, because the average inventory level equals CEP / 2 plus the safety stock. Again, there is a balance between costs, but the optimal inventory policy is still offering the lowest total cost.

Point calculation stocks replenishment and security:

POINT refill (1)	EXISTENCE OF SECURITY (2)	AVERAGE INVENTORY (3)	COST DEPLETION (4)	MAINTENANCE COST (5)	COST refill (6)	TOTAL COST (7)
		Q / 2 + (2)		$C\left(\frac{Q}{2}\right)$	$F\left(\frac{S}{Q}\right)$	(4) + (5) + (6)

Point Refills = Average demand during lead time + safety stock

average inventory = Q / 2 + safety stock

Costs exhaustions are estimated for each level management safety stock.

ABC INVENTORY CONTROL SYSTEM

ABC inventory control system. Classifies inventory into one of three groups-A, B or C based on the critical needs. The most important items are A, and the less important are the C. For example, suppose 10% of inventory items representing 80% of the total inventory. These could be the articles A. The articles B may represent 30% of the total number of items but only 15% of the total inventory. Items C represent 60% of the number of items but only 5% of the total inventory.

Being the most critical items to be administered very carefully. Articles B are not administered so carefully, and the articles C are administered with less care of all. For example, for an automobile factory, engines, transmissions and indeed all components of automobiles that are built are items A and administered with care. On the other hand, office supplies like paper clips, pencils and paper are ordered when needed without much supervision.

MRP SYSTEMS SUBJECT (PNM)

Companies often produce or provide more than one product, each product could have a large number of components. Usually, this means that the company has multiple suppliers. Coordination and programming can become very complex and that is why companies rely on systems Material Requirements Planning (PNM). The PNM systems are computerized systems that plan backward from the production schedule for purchases and inventory control. These large software systems combine information about the production process and procurement process to determine when the company must carry out orders. If done correctly, the PNM ensures that production will continue without interruption due to stockouts.

INVENTORY SYSTEMS JUST IN TIME (JIT)

Inventory systems just in time (JIT) significantly reduce inventories. JIT philosophy of a system is that the materials should arrive just at the time they are needed in the production process. The system requires careful planning and scheduling, and extensive cooperation between suppliers and manufacturers throughout the production process. All this is facilitated by electronic data interchange. A JIT system can reduce inventories of raw materials, production process and finished product.

JIT success of a system depends on several factors.

Planning needs. JIT requires a coordinated and integrated company-wide plan. Remember that one of the basic functions of inventory is to serve as a cushion in different stages of the production process. Through careful planning and scheduling, JAT mattress systems virtually eliminate these stocks. JIT integrated operating environment can produce substantial savings, whether for a business not practical to achieve a high degree of coordination and planning, JIT will not work.

Supplier Relations. JIT to work, the company must work closely with its suppliers. Delivery schedules, quantities, quality and instant communication are key ingredients of the system. The system requires frequent deliveries of the exact quantities needed and the order in which they are required. One must check the supplies carefully, often with a barcode. Therefore, it is essential to maintain good relations with suppliers.

Preparation costs. The manufacturing process is redesigned so that it is as flexible as possible, reducing the length of production runs, it is common in factories have fixed cost of preparing every time a production run begins. The optimal size of series production depends on the cost of preparing (similar to how inventory restocking costs depend on the fixed cost of ordering). If the company reduces these costs of preparation, one may have shorter production runs that will confer greater flexibility.

Other cost factors. Since JAT systems require careful monitoring and control companies using JAT generally they limit the number of its suppliers to reduce their costs. Many companies can reduce their maintenance costs and inventory preparation using a JIT system, but nothing is free. It must require suppliers higher quality, more frequent deliveries, sequential shipments and marked with bar codes, etc. Additional management costs from suppliers raise their prices. However, the balance has been very profitable, and many companies have adopted the JIT philosophy.

Impact on credit conditions. The JAT systems would be impossible without electronic data interchange. Since many aspects of the procurement process-production-marketing is now handled electronically, trade credit is also automated. When electronic credit conditions are used, the payment is not made, say, 30 days after the date of invoice. Rather, payment can be made soon after (maybe one day) of delivery and use of materials. Basically, this eliminates the accounts payable of a company that was a major source of short-term financing. On the other hand, electronic billing eliminates vendor accounts receivable and also eliminates the costs and risks associated systems payables and receivables based on paper.

ACCOUNTS RECEIVABLE MANAGEMENT

Credit sales receivables created few. There are two types of credit Commercial Credit and Consumer Credit. There is a commercial credit when a company buys goods or services to others without paying them simultaneously. Such sales receivable create a sale for the supplier (seller) and how much to pay for the purchasing company. Consumer credit, or retail credit is created when a company sells goods and services to a consumer without a simultaneous payment.

Most business transactions using commercial credit. In the retail level, payment mechanisms include cash, checks, credit extended by the retailer and credit extended by a third party (such as Master Card, Visa or American Express). If we ask because companies provide credit, managers often answer that they must do so because their competitors do. Although true, this response is simplistic and does not touch the real reasons for the widespread use of credit.

Trade credit is actually a loan from one company to another, but a loan that is linked to a purchase. By using commercial credit, both sides of the transaction should be able to reduce the cost and risk of doing business. Some considerations specific market.

Financial intermediation. Overall, the interest rate of a loan commercial loan benefits both parties: it is lower than the other rates that could borrow the client, but higher rates of short-term investment that could get the supplier.

Warranty. Suppliers know how to use the goods as better than other providers, such as banks guarantee.

Information costs. A provider may already possess the information necessary to determine whether a customer is credit worthy or not. If a bank wants to lend money to the same customer, the lending decision entails costs. This gives the supplier an advantage over the bank, it does not have to incur such costs.

Information Quality Product. Providers generally have better information than customers about the quality of their products. If the product has acceptable quality, commercial credit customer pays on time. If the product is of low quality, customer returns and refuses to pay.

Opportunism Employees. Companies try to protect themselves against theft by employees in various ways. One is to separate employees who authorize transactions, those who physically handle the products and those who manage payments.

Steps in the distribution process. By giving credit to the ultimate purchaser, the payment mechanism ignores all actors in the distribution process and requires only payment of the final buyer to the original seller.

Safety and comfort Psychology Buyer. Sometimes it is not convenient to pay at the time of purchase. Carry a lot of cash increases the possibility of being victims of theft and the possibility of lost or forgotten cash leave, psychology is important, especially at the level of detail sale

THE DECISION GRANTING OF BASIC CREDIT

The basic analysis for lending decisions is the same as for other financial decisions. Credit should be given where such action is a decision with Net Present Value (NPV) positive.

NPV = VP of future net cash inflows - exit

In the case of a simple decision to grant credit, the NPV is:

$$VPN = \frac{pR}{(1 + k)^t} - C \tag{1}$$

At zero time, we invest C in a credit sale. The investment could be the cost of goods sold plus sales commissions. The sale amount is R The probability of payment is p, and the expected payoff is pR. The probability of customer payment is estimated subjectively or with the help of statistical models. Payment is expected at time t. The required performance is k. If the NPV is negative, it should not be granted credit. What we want is a positive VPN. We can also calculate a probability payment indifference (zero VPN), p *.

If a customer credit has a higher probability of payment p *, granting credit has a positive VPN. This probability is calculated indifference payment equaling VPN zero in the above equation (1) and solving for p:

$$p = \frac{C\,(1 + k)^t}{R} \tag{2}$$

EXAMPLE: A company has a customer who wants to buy \$ 1,000.00 merchandise on credit. It is estimated that there is a 95% probability that the customer will pay \$ 1,000.00 in three months and a probability of 5% of total default (not pay anything). Suppose an investment amount of 80%, made at the time of sale, and a required 20% yield. What VPN is the granting of credit, and what is the probability of payment of indifference?

Using equation (1), the VPN is:

$$VPN = \frac{pR}{(1 + k)^t} - C$$

Therefore, granting credit is profitable, and is recommended. Using equation (2), the probability of indifference payment is:

$$p = \frac{C(1 + k)^t}{R}$$

This result is consistent with the calculation of the VPN. The estimated probability of 95% paid by the company exceeds the value of indifference and indicates that credit should be given.

SOURCES OF CREDITS

There are several valuable sources of internal and external information. Primary insiders are:

A credit application, including references.

Payment history of the applicant, if it has extended credits previously.

Information sales representatives and other employees.

External sources of credit information are important:

Financial statements of recent years. The financial statements can be analyzed to assess the profitability, liquidity and debt obligations of the customer.

Reports from credit rating agencies such as Dun & Bradstreet Business Credit Services (D & B). These agencies provide credit assessments of thousands of company and estimates its overall soundness. D & B assesses the credit of a company in relation to other comparable companies with financial strength and credit ratings assigned integrated assessment of 1 (“High”) and 4 (“Limited”), in the United States.

Reports credit bureaus. These reports provide objective information about financial obligations due from a company. Reports from credit bureaus also contain information on any legal opinions against the company.

Credit files of industry associations. Industry associations often maintain credit files. These partnerships and direct competitors one are often willing to share credit information about customers.

THE FIVE C CREDIT

The five Cs of credit are five general factors credit analysts often consider when making a decision to grant credit.

Character. Commitment to fulfill credit obligations. The best way to measure character is based on the applicant's payment history.

Capacity. The ability to meet credit obligations with current income. The ability is assessed by examining the cash inflows of the income statement or cash flow statement of the applicant.

Capital. The ability to meet credit obligations using existing assets if necessary. Capital is evaluated by looking at the net value of the applicant.

Collateral (Warranty). The guarantee can be disposed of in the event of nonpayment. The value of collateral depends on the cost of disposal and the potential resale value.

Conditions (general economic or industry conditions). External customer's business conditions affecting the decision to grant credit. For example, an improvement or a deterioration in general economic conditions may change interest rates or the risk of extending credit. Furthermore, the conditions of a particular industry may affect the profitability of lending to a company in that industry.

ACCUMULATION ACCOUNTS RECEIVABLE

The total amount of receivables outstanding at any given time is determined as a product of two factors:

The volume of credit sales and the average time between sales and collection. If they change credit sales or collection period, such changes will be reflected in the accounts receivable. When a credit sale is made, the following things happen:

Inventories are reduced by the cost of goods sold.

Accounts receivable increased by the sale price, and

The difference gain, which is added to the utilities.

If the sale is cash, then the company has in fact received the cash from the sale, but the sale is on credit, the company will not receive cash from the sale until the account is charged.

Collection policy

The company determines its overall policy collection by combining recovery procedures undertaken. These procedures include customer a phone call, sending a letter, sending a new bill, sometimes to visit a person and legal action.

$$Average\ collection\ period = \frac{accounts\ receivable}{annual\ credit\ sales / 360} = \frac{accounts\ receivable}{sales\ per\ day}$$

$$Average\ accounts\ receivable = \frac{annual\ credit\ sales}{360 / average\ collection\ period}$$

SHORT-TERM FINANCING.

The short-term credit is defined as debt originally scheduled to be repaid within one year. There are two common methods to guarantee loans:

Financing Accounts Receivable

Inventory financing.

Commercial credit

In the ordinary course of events, a company buys its supplies to other companies and materials on credit and debt recorded as an account payable. Accounts payable, or trade credit, constitute the largest single category of short-term credit, accounting for about one third of the current liabilities of non-financial corporations. This percentage is larger for small businesses so much. Because these companies can not qualify for financing from other sources, they are based primarily on trade credit. The four main factors that influence the length of credit terms:

1. Economic nature of the product. Satisfiers with a high turnover of sales sold in terms of shorter or longer credit; buyers resell the products quickly, and thus generate cash enables them to pay suppliers.
2. Circumstances of the seller. Financially weak sellers must require cash or credit terms exceptionally short. In many industries, changes in credit terms can be used as a sales promotion.

   Large sellers can use their position to impose relatively short terms, although in practice the opposite occurs; ie financially strong sellers are providers of funds for small businesses.
3. Circumstances Buyer. Generally, retail traders financially sound, selling on credit, can both receive slightly longer terms

of credit. Some menuditas, such as those sold in particularly risky areas (such as clothing) receive, but extended credit terms usually be offered deep discounts to encourage prompt payment.

4. Cash Discount. It is the reduction in price is based on a payment within a specified period. The costs of not taking cash discounts exceed the interest rate at which the buyer can borrow; It is therefore important that a company be cautious when using trade credit as a source of funding: could be very costly. The actual duration of the credit as modified by the magnitude of the discounts offered.

Net Credit concept

Commercial credit has a double meaning for the company. It is a source of credit to finance purchases, and is an application of funds to the extent that the company finance customers credit sales.

Large well-funded companies and businesses of all sizes tend to be net trade credit providers; small and undercapitalized of all sizes tend to be net users of commercial credit. It is impossible to generalize whether to a net supplier or net user; The choice depends on the circumstances of the company and the various costs and benefits of receiving and using trade credit.

Advantages of Commercial Credit as a funding source:

Commercial loans, a customary way of doing business in many industries, it is convenient and casual. A company that does not qualify for credit from a financial institution can receive trade credit because past experience and familiarized the seller with dignity customer credit. The seller knows the marketing practices of industry and usually has a good position to judge the ability of the customer and the risk of selling on credit. The amount of trade credit purchases fluctuates in accordance with credit limits that may be operating.

Importance of Good Relations Provider during inflation

During periods of inflation and little money, companies increase their standards to extend trade credit to their customers. The settlement of accounts receivable is a way to get a more favorable liquidity position.

Therefore, providers become more selective when extending trade credit. This demonstrates the importance of a business the importance of its suppliers win. Have good financial reasons and pay quickly are excellent ways to achieve this goal. But even if these indicators are unfavorable, a company may be able to obtain commercial credit offering realistic plans to improve their situation.

Short-term financing through commercial banks

Loans from commercial banks, which appear in the balance sheet as notes payable, rank second in importance of trade credit as a source of short-term financing. Banks occupy a basic position in the money markets short and intermediate term. His influence is greater than it appears from the peso amounts lending because banks provide funds not spontaneous. As needs increase financing a business, this calls for additional funds to banks. If the request is denied, the alternative is to decrease the growth rate or reduce operations.

Characteristics of loans from commercial banks.

Loan forms. One gets a business loan from a bank is in principle no different from a loan taken out by an individual. In fact, it is often difficult to distinguish between a bank loan to a small business and a personal loan. The loan is obtained by signing a conventional promissory; reimbursement, with an accumulated sum at maturity (when the document expires) or in installments over the life of the loan.

Customer size. Banks make loans to any size business. By amount in pesos, large companies get most of the bank loans. But number of loans to small businesses it is up more than half of bank loans.

Expiration. Commercial banks are concentrated in the market for short-term loans. Short-term loans make up about two-thirds of bank loans by dollar amount, while installment loans (loans with maturities over one year) barely one third (USA).

Warranty. If a potential borrower is a questionable credit risk, or if the funding needs of the company exceed the amount that the officer considers it prudent bank loans on an unsecured basis, it will require some form of collateral

Reciprocity. Banks require regular borrower to maintain an average balance in the checking account. These balances, commonly called reciprocity, are a method of raising the real rate of interest.

Repayment of bank loans.

Almost all bank deposits are subject to withdrawal on demand; why commercial banks try to prevent companies from using a bank loan to permanent financing. A bank therefore may require their borrowers "clean" short-term bank loans at least one month each year. If a company can not be freed from their bank debts at least part of the year, it is using bank financing for their ongoing needs and should seek other sources of permanent financing or long term.

Cost of commercial bank loans.

Almost all commercial bank loans have recently cost between 10 and 20% depending on the actual rate of company characteristics and level of interest rates in the economy. If the company qualifies as preferential risk due to their size and financial strength, the interest rate will be one half to three quarters of 1% on the discount rate charged by Federal Reserve banks to commercial banks. This is called prime rate. Moreover, with a small financial reasons that do not exceed the average company, one may be required to provide collateral and pay a real interest rate of 2 to 3 points above the prime rate.

"Regular" Interest

Determining the real rate of interest on a loan depends on the stipulated rate of interest and the method applied by the lender to charge interest. If interest is paid at maturity of the loan, the interest rate provided is the real rate of interest. For example, on a loan of \$ 20,000.00 to one year 10% interest is \$ 2,000.00

$$Regular\ loan\ interest\ paid\ at\ maturity = \frac{Interest}{Loan}$$

$$= \frac{\$\,2{,}000.00}{20{,}000.00} = 10\%$$

Offline interest

If the bank deducts the interest in advance (deducts the loan), increase the real rate of interest. On the loan of $ 20,000.00 a year, 10% discount is $ 2,000.00 and the borrower obtains the use of only $ 18,000.00 real interest rate is 11.1% (and not 10% as it would be on a "regular" loan)

$$Discounted\ loan = \frac{Interest}{loan\ request - interest}$$

$$= \frac{\$\,2{,}000.00}{\$18{,}000.00} = 11.1\%$$

Loan installments

Loans "regular" interest and discounted set forth above are characterized by the repayment of the loan principal at maturity. According to the method in installments, principal payments are periodically (eg every month) Over the term of the loan; On a one-year loan, the borrower has the full amount of money just for the first month, and last month he has already paid eleven twelfths of the loan. Therefore, the actual rate (ie, the annual percentage rate, or APR annual percentage rate) of interest on the loan in installments is significantly higher than the stipulated rate. Installment loans can be arranged in two ways. In a fertilizer additive loan, the net proceeds of the loan are the same as the nominal value of the loan, but the interest is added to the loan principal to calculate monthly installments. In a discounted loan installments, interest is subtracted from the principal for the net proceeds of the loan while fertilizers are based on aggregate face value of the loan.

The first type of loan in installments results in higher monthly payments but a real rate of interest lower; The second type of loan in installments results in lower monthly payments but higher real rate of interest.

Optimization of bank products

Banks have direct relationships with their borrowers. It is established over the years a lot of personal association, and often commented upon business problems lender. Thus banks provide informal advice administrative services. A lender seeking such a relationship should know the important differences that exist between banks, and then consider:

Banks have different basic policies toward risk. Some are fairly conservative lending practices; others participate in what has been called creative banking practices.

Some officials of bank loans are usually requested in the provision of services and to stimulate development loans to companies in their early and formative years.

Banks differ in the extent to support the activities of borrowers in bad times. This feature is called degree of loyalty of the bank

Another feature in which differ banks is the degree of stability of the deposits. Instability not only arises from fluctuations in the level of deposits but also its composition.

Banks differ in thc degree of specialization of loans. Larger have departments that specialize in different types of loans; Real estate loans in commercial fertilizers.

The size of a bank can be an important feature. The maximum loan that can be a bank to any customer is generally limited to 10% of the few bank's capital (capital plus retained earnings), and generally is not appropriate that large companies hold lending relationships with small banks.

With increasing competition among commercial banks and other financial institutions, it has increased the aggressiveness of banks. Modern commercial banks now offer a wide range of financial services and business.

COST OF CAPITAL AND ITS IMPLICATIONS FROM BUSINESS INVESTMENT.

Investments made by the company are fundamental for the development of the corporate purpose of the organization, the cost of capital emerges as a relevant factor in assessing the financial manager makes investment projects that have short-term

CAPITAL COST

The cost of capital is the rate of return one should get the company on its investments to their market value remains unchanged, considering that this cost is the discount rate of future corporate profits, is why the corporate finance manager should be provided with the necessary tools to make decisions on investments to be made and therefore the most convenient to the organization.

The study of the cost of capital is based on the specific sources of capital to seek the fundamental inputs to determine the total capital cost of the company, these sources should be long term, as these are the ones that provide permanent funding.

The main sources of long-term funds are long-term debt, preferred stock, common stock and retained earnings, each associated with a specific cost and leading to the consolidation of the total cost of capital.

Fundamental factors underlying capital cost are the following:

The degree of business and financial risk.

Tax and tax impositions.

Supply and demand for funding resources

COSTS OF SOURCES OF CAPITAL

Here is a small theoretical approach of the costs of these sources is presented, but keep in mind that the study of each of these costs has deeper connotations in its application.

Cost of long-term debt

This source of funds has two primary components, the annual interest and amortization of discounts and premiums to be received when the debt was incurred. The cost of borrowing can be found by determining the internal rate of return of the cash flows related to debt. This cost represents for the company the annual percentage cost of debt before taxes.

Cost of Preferred Stock

The cost of preferred stock is found by dividing the annual dividend preference share, from the net proceeds of the sale of the preferred stock.

CAP = Annual dividend per share / Net proceeds on sale of AP

Cost of common shares

The value of the cost of common stock is a bit complicated to calculate, since the value of these shares is based on the present value of all future dividends that will be paid on each share. The rate at which future dividends are discounted to present value to turn represents the cost of the common shares.

Cost of retained earnings

The cost of retained earnings is closely linked to the cost of the common stock, if not profits these would be paid to common shareholders as dividends were retained, it is then that the cost of retained earnings is considered the opportunity cost of dividends transferred to existing common shareholders.

COST OF CAPITAL

In determining the costs of specific sources of long term financing, the technique usually used to determine the total cost of capital used in the financial evaluation of future company investments shown.

The main method to determine the total and appropriate cost of capital is to find the average capital cost basis using historical or marginal costs. The financial manager must use the capital cost analysis to accept or reject investments, because they are what define compliance with organizational objectives

AVERAGE COST OF CAPITAL

The average cost of capital is weighing the cost of each specific type of capital or marginal historical capital of each type is used proportions. Historical weights are based on the existing capital structure of the company, while the marginal weights consider the actual proportions of each type of financing is expected to finance a given project.

The use of historical weights to calculate the average cost of capital is quite common, they are based on the assumption that the existing composition of funds, or capital structure is optimal and therefore should be held in the future. One can use two types of historical weights:

Using book value weights means that new funding is achieved using exactly the same proportion of each type of financing that the company currently has in its capital structure.

Weightings of market value

Financiers for this is more attractive than the previous one, because market values values are closer to the actual amount received by selling them. Moreover, as the costs of different types of capital are calculated using prevailing market prices, it seems reasonable also use weights market value, however, is more difficult to calculate the values in the market sources equity financing of a company to use book value.

The average cost of capital based on weightings market value is usually higher than the average cost-based weights book value, since most of the preferred shares and common have market values much higher than the value in books.

Using marginal weights

The use of marginal weights involves specific cost weighting of different types of financing by the percentage of total funding that are expected to arise with each method of historical weights. By using marginal weights it refers primarily to the actual amounts of each type of financing used.

With this type of weighting one have a real process of project financing and admits that the funds actually are available in different quantities, using different sources of long term financing also reflects the fact that the company does not have much control over the amount financing obtained by the surplus

One criticism that is made using this system is that it does not consider the long-term implications of the current funding of the company.

The formula for the cost of the capital is given here:

$$CC = \frac{E}{(D + E)} r_e + \frac{D}{(D + E)} (1 - T)\, r_d = (1 - L) r_e + L(1 - T)\, r_d$$

where E is market value of capital, D denotes market value of debt and re is the performance required capital. The following is valid:

$$r_e = \frac{D_1}{p_0} + g$$

where D1 is value of expected future cash dividend, P0 is fair price of one common share (current market value of common shares), g is yield (growth rate), T is tax rate, rd is debt performance required.

L is ratio between debt financing and the total investment value given by

L = D / (D + E)

CHAPTER 7

PERSONALITY TRAITS AS PREDICTORS OF SHOPPING MOTIVATIONS AND BEHAVIORS: A CANONICAL CORRELATION ANALYSIS

Ali Gohary [a], Kambiz Heidarzadeh Hanzaee [b]

[a] Department of Business Administration, Qazvin Branch, Islamic Azad University, Qazvin, Iran

[b] Department of Business Management, Science and Research Branch, Islamic Azad University, Tehran, Iran.

ABSTRACT

This study examines the relationship between Big Five personality traits with shopping motivation variables consisting of compulsive and impulsive buying, hedonic and utilitarian shopping values. Two hundred forty seven college students were recruited to participate in this research. Bivariate correlation demonstrates an overlap between personality traits; consequently, canonical correlation was performed to prevent this phenomenon. The results of multiple regression analysis suggested conscientiousness, neuroticism and openness as predictors of compulsive buying, impulsive buying and utilitarian shopping values. In addition, the results showed significant differences between males and females on conscientiousness, neuroticism, openness, compulsive buying and hedonic shopping value. Besides, using hierarchical regression analysis, we examined sex as moderator between Big Five personality traits and shopping variables, but we didn't find sufficient evidence to prove it

INTRODUCTION

Personality traits are one of the main sources of our decisions. Furthermore, social personality can predict the social behaviour in particular conditions. In this paper, we don't want to examine what factors impacts on personality, but we want to explore the personality effects on shopping motivations to predict future behaviour of our new or current customers and illustrate a guideline map to plan and conduct our strategic programs. A Study on the impact of personality on shopping section will modify our approach to the business; what goods and services should we produce? What is the customer's response to specific social stimuli that have impact on personality and what proper reaction should we do? Which personality trait has outstanding role in the society so that we will provide suitable goods and services to be alive in this emerging market? Are men and women same in shopping procedure? And when do they notice hedonistic or utilitarian aspects of shopping? In other words, individuals have different types of personality traits which are bold within their personality and make them distinctive in behaviour, habits, motivations and responses to a stimulus. Individual differences in personality come from two sources: environmental sources which are early experiential calibrated, enduring situational evocation, strategic specialization, adaptive self-assessment of inheritable qualities and heritable sources, which are temporal or spatial variations in selection pressures, negative frequency-dependent selection and mutation–selection balance (Buss, 2008). For several centuries, psychologists tried to determine the dimensions and characteristics of personality. These efforts had continued until they increasingly agreed that five supertraits might adequately describe the structure of personality (Maltby, Day, & Macaskill, 2010). Costa and McCrae (1985, 1992a, 1992b, 1995) discussed five major factors, which are influential in forming personality.

They called these traits as Big Five Model of Personality and developed a measure to assess them that is called NEO Personality Inventory. Five dimensions of Big Five Factors are Neuroticism, Extraversion, Openness (to experience), Conscientiousness, Agreeableness (Digman, 2002). These personality factors were conceived by researchers to pay much more attention to relations between personality and consumer behaviour. Although personality research ("personology") has long been a fringe player in the study of consumer behaviour, little research has directly been devoted to personality issues, and if consumer personality has ever been investigated, it tended to be from the narrow perspective of developing yet another individual difference measure in an already crowded field of personality scales or considering the moderating effects of a given trait on some relationship of interest (Baumgartner, 2002). But recently, many researchers discussed the topic of personality and consumer behaviour (Verplanken & Herabadi, 2001; Wang & Yang, 2008).

The importance of investigating shopping behaviour in Iran

In this study, we examine personality and shopping variables among a sample that was chosen from Iranian college students. The main reason that we chose Iran was the rapid economic and social changes in the last years that have direct impact on society's social behaviour; for example, based on IMF (International Monetary Fund), Iran's GDP was increased and wealth dissemination index got a better rate, the Gini ratio improved and women contribution rate ascended rapidly. These factors caused to raise a new category of mid-income families with higher education, different needs, and unfamiliar with urban living challenges might affect in their personality. In addition, Iran's population increased very fast from 1979 to 1990. The birth rate reached over 3% per year. Total population goes higher from 35 million up to more than 65 million. In this era, a new Iranian Generation Y was introduced regarding those who prefer to live in urban regions and immigrated therefore from rural to urban regions in the last years for better job opportunities and educational purposes. These factors drove people to create their new identity throughout consumption. On the other side, by increasing oil price which is Iran's major income source, demand for consuming (luxury) goods especially imported goods increased, so that imports set a record and go over 58 billion dollars in

2011. By spread of materialism in Iran in these years (Teimourpour & Hanzaee, 2014), social shopping behaviour and social personality were impressive. In this context, most of international foreign companies such as Apple, HP, Hyundai, L'Oréal, Porsche, Knor, Procter & Gamble, and Siemens started to export their goods and services to Iran and open a representation without paying attention to sanctions. Others like Nestle, Unilever, Samsung, LG, Volvo directly manufactured their product in Iran. Meanwhile, the lack of enough researches related to shopping and psychology in this market makes of this research a necessary study for awareness-raising Iranian shopping behaviour.

Literature Review and Hypotheses

Research suggests that the shopping experience provides consumers with a combination of utilitarian and hedonic shopping value (Carpenter & Moore, 2009; Babin, Darden, & Griffin, 1994), impulsive and compulsive buying. Considering five major personal traits that we wanted to test which ones drove shopping motivations, a concise literature review of this factor has been described hereunder. At first, we explained four major shopping variables and then, personality traits will be provided to explore theoretical linkage among personality and shopping variables.

Impulsive Buying: Impulse buying generates over $4 billion in annual sales volume in the United States. With the growth of e-commerce and television shopping channels, consumers have easy access to impulse purchasing opportunities, but little is known about this sudden, compelling, hedonically complex purchasing behavior in non-Western cultures (Kacen & Lee, 2002). Impulsive buying has been defined as the spontaneous or sudden desire to buy something, and when compared to more contemplative approaches to decision-making, it considered emotional, reactive, and "prone to occur with diminished regard" for the consequences (Rook, 1987). Rook (1987) also stated: "Impulse buying occurs when a consumer experiences a sudden, often powerful and persistent urge to buy something immediately. The impulse to buy is hedonically complex and may stimulate emotional conflict. Also, impulse buying is prone to occur with diminished regard for its consequences". Recent research on impulse buying behavior indicated that individual consumers did not view their specific purchases as wrong and indeed retrospectively reported a favorable evaluation of their behavior (Hausman, 2000).

Compulsive Buying: Faber and O'Guinn (1988) defined compulsive consumers as "people who are impulsively driven to consume, cannot control this behavior, and seem to buy in order to escape from other problems" (Mowen & Spears, 1999). DeSarbo and Edwards (1996) concluded that compulsive consumption was related to some of psychological traits such as "dependence, denial, depression, lack of impulsive control, low self-esteem, approval seeking, anxiety, escape coping tendencies, general compulsiveness, materialism (envy), isolation, excitement seeking, and perfectionism". Most research and scholars consider that excessive buying, defined as consumer spends more than he/she can afford or beyond his/her needs, is responsible for this situation. The results of this painful issue for individuals, families, societies as well as countries and business environment are all unfavourable. For this reason, this behaviour has been of theoretical and practical interest to psychologists, psychiatrists, economists, sociologists and marketing scholars and practitioners (Eren, Eroglu, & Hacioglu, 2012). That is why, authors discuss compulsive buying in terms of personality traits, demographic variables, family structure and the patterns that classify buyers as compulsive and non- compulsive (Faber & O'Guinn, 1992;Faber, O'Guinn, & Krych, 1987;Faber & O'Guinn, 1989).

Hedonic Shopping Value: Hirschman and Holbrook (1982) defined hedonic consumption as "those facets of consumer behavior that relate to the multisensory, fantasy and emotive aspects of one's experience with products." Basically this values are unstructured, mentally, affective and experience-based (Kim & Han, 2011) stimuli driven, pleasant and fun (Nguyen, Nguyen, & Barrett, 2007). People with higher hedonic values cannot be satisfied with utilitarian or functional aspects of buying behaviors but enjoyable and pleasurable aspects of them and they regard emotional and psychological values of shopping experience. Hedonic values are assumed to be associated with gratification through fun, fantasy, playfulness and enjoyment (Eren, Eroglu, & Hacioglu, 2012). Hedonic value derived from the shopping experience reflects the emotional or psychological worth of the purchase. Sources of hedonic value could include the joy and/or the excitement of shopping, or the escape from everyday activities that is provided by the experience (Carpenter & Moore, 2009).

Utilitarian shopping values: The utilitarian perspective is based on the assumption that consumers are rational problem-solvers. As a result, the utilitarian perspective stresses functional and product-centric thinking, and the research has focused on consumer decision processes. Consumption is understood as a mean to accomplish some predefined end (Rintamäki, Kanto, Kuusela, & Spence, 2006). Consumers perceive utilitarian value by acquiring the product that necessitated the shopping trip while simultaneously perceiving hedonic value associated with the enjoyment of the shopping experience itself. Utilitarian value reflects shopping with a work mentality (Hirschman & Holbrook, 1982). Consumers seek utilitarian value in a task-oriented, rational manner (Carpenter & Moore, 2009). Utilitarian consumer behavior has been described as ergic, taskrelated, and rational (Babin, Darden, & Griffin, 1994).

We characterized Extraversion as the dimension underlying a broad group of traits, including sociability, activity, and the tendency to experience positive emotions such as joy and pleasure (Costa, Jr & McCrae, 1992). Introversion and social isolation are opposite of extraversion, which is one of the Big Five Factors (Mowen & Spears, 1999). In a research about shopping experiences, Guido et al. (2007) stated that when individuals were introverted, their shopping behaviour tended to be utilitarian; whereas, when the same enduring trait of personality was directed towards extroversion, their shopping behaviour tended to be hedonic. In addition, Matzler et al. (2006) found that extraversion was positively related to the hedonistic value of the products.

Therefore, it sought to investigate this match-up hypothesis as well.

H_{a1}: There is an indirect relationship between extraversion and impulsive buying.

H_{a2}: There is a direct relationship between extraversion and compulsive buying.

H_{a3}: There is a direct relationship between extraversion and Hedonic shopping values.

H_{a4}: There is an indirect relationship between extraversion and Utilitarian shopping values.

High Openness (to experience) individuals are imaginative and sensitive to art and beauty and have a rich and complex emotional life; they are intellectually curious, behaviourally flexible, and non-dogmatic in their attitudes and values (Costa, Jr & McCrae, 1992). They are not conventional in their ideas, values, and beliefs. Low open to experience people are conventional and present narrow interests (Lakhal, Frenette, Sévigny, & Khechine, 2012). Matzler et al. (2006) concluded that the higher open individuals tended to be curious about both inner and outer worlds, to have experientially richer lives, and to experience both negative and positive emotions more keenly than closed individuals. It can be assumed that they perceive and experience hedonic values of products stronger than individuals who score low on openness. Some researches confirmed the relationship between openness and intelligence (DeYoung, Peterson, & Higgins, 2005; Zurawicki, 2010). It attempts to determine the biological conditioning of intelligence can prove important to identify the problem-solving skills and, hence, the decision making patterns (Zurawicki, 2010) that have a crucial role in predicting shopping behaviours and motivations. It is noticeable that Voss et al. (2003) stated problem solving as a subcategory of utilitarian values. Then,

H_{a1}: There is a direct relationship between openness and impulsive buying.

H_{b2}: There is a direct relationship between openness and compulsive buying.

H_{c3}: There is a direct relationship between openness and Hedonic shopping values.

H_{d4}: There is an indirect relationship between openness and Utilitarian shopping values.

Neuroticism represents the individual's tendency to experience psychological distress, and high standing on neurotic is a feature of most psychiatric conditions (Costa, Jr & McCrae, 1992). In relation to Neuroticism, people high on N tend to be emotionally labile and frequently complain of worry and anxiety as well as of bodily aches (e.g., headaches, stomach difficulties, dizzy spells), they are hypochondriac, insecure, and inadequate (Pervin, 2006). The opposite of neuroticism is emotional stability (Goldberg, 1990). Prior researches reported the relationship between compulsive buying and neuroticism. Johnson and Attmann (2009) found a significant relationship between compulsive buying and neuroticism. However, according to Mowen et al. (1999) individuals who got a high score on neuroticism were compulsive buyers. In addition, d'Astous et al. (1990) found a negative relationship between self-esteem and compulsive buying. It is noticeable Costa, Jr & McCrae (1995) concluded that by increasing the level of neuroticism, individuals tending to impulsiveness, depression, anxiety and vulnerability would be more intensive. Additionally, Mick (1996) found a positive relationship between impulsive buying and neuroticism. Chetthamrongchai and Davies (2000) suggested about individuals who scored relatively high on present orientation which was indicating that they were more concerned with that was happening at the moment than in the past or in the future. In addition Roberts and Pirog (2004) found a positive link between extrinsic goals, which were associated with low self-esteem, and compulsive buying. So,

H_{3a}: There is a direct relationship between neuroticism and impulsive buying.

H_{3b}: There is a direct relationship between neuroticism and compulsive buying.

H_{3c}: There is an indirect relationship between extraversion and Hedonic shopping values.

H_{3d}: There is a direct relationship between neuroticism and Utilitarian shopping values.

Agreeableness is an expression of the need for harmonious relations, which implies the rejection of the domineering approach (Zurawicki, 2010). High agreeable individuals are trusting, sympathetic, cooperative, good natured, straightforward, forgiving, and gullible; low-An individual are cynical, callous, and antagonistic (Costa, Jr & McCrae, 1992; Pervin, 2006). These people with higher scores on agreeableness tend to trust rather than to be suspicious of other people. In other words, individuals with a lower degree of agreeableness doubt things that they do not personally know (Wang & Yang, 2008). Ho et al. (2004) suggested that agreeable people would not experience as strong a negative emotional response, as less agreeable people, and these people are better at emotional self-regulation, including the regulation of anger and other negative emotions. Duijsens and Diekstra (1996) reported a negative relationship between agreeableness and compulsive and impulsive behavior, but Balabanis (2002) and Wang et al. (2008) didn't find sufficient evidence to support this claim. Guido et al. (2006) suggested agreeable people would have a strong linkage with hedonic shopping values. Then we hypothesis:

H_{4a}: There is an indirect relationship between agreeableness and impulsive buying.

H_{4a}: There is an indirect relationship between agreeableness and compulsive buying.

H_{4a}: There is a direct relationship between agreeableness and Hedonic shopping values.

H_{4a}: There is a direct relationship between agreeableness and Utilitarian shopping values.

Conscientiousness is a dimension that contrasts scrupulous, wellorganized, and diligent people with lax, disorganized, and lackadaisical individuals (Costa, Jr & McCrae, 1992). These people are reliable, selfdisciplined, punctual, neat, ambitious, persevering, deliberating, competent, dutiful (Pervin, 2006; Maltby et al., 2010), who display planned rather than spontaneous behavior and at the extreme, like the individuals tending to be perfectionists (Zurawicki, 2010). Conscientiousness people are able to control impulse emotions and delay gratification (Joshanloo, Rastegar, & Bakhshi, 2012). This ability drives them to not be an impulse shopper as the reports from Gustavsson et al. (2003) showed a negative relationship between impulsive behavior and conscientiousness. Mowen et al. (1999) in a research found a significant negative relationship between compulsive buying and conscientiousness and suggested that individuals who had difficulty controling their buying might also reveal a lack of organization, precision, and efficiency in their daily lives. Therefore, the following hypotheses are proposed:

H_{5a}: There is a direct relationship between Conscientiousness and impulsive buying.

H_{5a}: There is a direct relationship between Conscientiousness and compulsive buying.

H_{5a}: There is a direct relationship between Conscientiousness and Hedonic shopping values.

H_{5a}: There is a direct relationship between Conscientiousness and Utilitarian shopping values.

Although we considered shopping as an activity to respond our daily needs, but nowadays it has an importance role in spending time and answers the different kinds of social pressures (which affect the person's behaviors) and lifestyle activity. Considering the different aspects of each personality trait, we want to answer the question that "Do personality traits predict shopping motivations?"

METHOD

Participants and procedure

To determine sample size, we used Statistical Power Analysis method which tried to make a balance between α (Error type I) and β (Error type II) to optimize hypothesis test and make the results more precious (Davey & Savla, 2010). So, we used GPower Ver 3.1 to calculate sample size (Faul, Erdfelder, Buchner, & Lang, 2009). The results of acquired samples showed the statistical power is 0.97, which was an acceptable amount to verify sampling procedure because statistical powers more than 0.6 are acceptable (Dattalo, 2008). Therefore, a self-report questionnaire was distributed randomly among 247 college students at Qazvin and Urmia in Iran. This sample consists of 149 male (60.3%) and 98 female (37.3%) students. Respondents were recruited from a variety of majors. A hundred-eighty-nine respondents were single and fifty-eight respondents were married. Average age of respondents was 25.53 years old (SD = 5.908) which 153 (63.22%) respondents were between 16- 25 years old, 68 (28.09%) respondents were 26-35 and 21 (8.69%) were 35-33 years old. Trained data collectors distributed questionnaires in the classrooms. They described the measures and answered the questions. Participants were convinced about confidentially of data, and they participated in the research voluntary.

Measures

The 74-item questionnaires were collected from samples. All the items of scales were translated to Persian by using the back-translation method. Due to the variety of variables of this research, we used different scales in our questionnaire. A shorten version of the NEO PI consisted of 44-items which was developed and obtained by John et al. (1991) and who was employed to measure Big Five personality traits in 2008. To assess Compulsive Buying, we used a scale which was developed by d'Astous et al. (1990). To measure Impulsive Buying, we employed a 9-item scale which was developed by Rook et al. (1995) in this research. Furthermore, we employed scales to assess Hedonic and Utilitarian Shopping Values that were developed by Babin et al. (1994) consisting of 11-items for hedonic shopping values and 4-items for utilitarian shopping values.

RESULTS

To assess the relationship between personality traits and shopping motivations, we analysed our data using by SPSS version 21. Preliminary analyses calculated bivariate correlations, means, standard deviations, and Cronbach's alpha are shown in Table 1. The results of bivariate analysis indicated that some of the relationships were supported recent studies and findings, but some of them were not significant.

Table 1: Intercorrelations between the scales and descriptive statistics among this study.

	1	2	3	4	5	6	7	8	9
1- Extraversion	1								
2- Agreeableness	.313**	1							
3- Conscientious-ness	.320**	.442**	1						
4- Neuroticism	-.384**	-.447**	-.372**	1					
5- Openness	.372**	.188**	.297**	-.279**	1				
6- Compulsive	-.012	-.055	-.164**	.124	.021	1			
7- Impulsive	-.027	-.042	-.132*	.136*	-.063	.720**	1		
8- Hedonic	.108	-.033	.015	.102	.017	.640**	.516**	1	
10- Utilitarian	.145*	.122	.266**	-.161*	.212**	-.018	-.113	.256**	1
Mean	3.22	3.71	3.47	2.90	3.57	2.78	2.39	3.06	3.50
Alpha .64 .63 .74 .71 .73 .73 .82 .87 .34									

* P<0.05; ** P<0.01;

Multiple regression

Four multiple regressions were performed to evaluate the role of Big Five personality as predictor of shopping motivations. We used stepwise method in procedure of multiple regression analysis, because it was the most appropriate way to determine the association between variables (Cramer, 2003). Table 2 shows the summaries of multiple regression results. The Outcome demonstrates that 2.7% of total variance in compulsive buying was explained by Big Five personality traits. This result suggested conscientiousness as significant predictor of compulsive buying (R^2= 0.027, R^2 (Adj) = 0.023, $F_{(245)}$ = 6.78, P<0.001) The results showed that 1.9% of total variance in impulsive buying was explained by Big Five personality traits (R^2= 0.019, R^2 (Adj) = 0.015, $F_{(245)}$ = 4.62, P<0.05). Also, the results suggested neuroticism as significant predictor of impulsive buying with standardized β coefficient of 0.136 (P<0.005). Third regression analysis didn't conclude on a certain trait as predictor of hedonic shopping values. Finally, the outcome of the last regression analysis demonstrates that 9% of total variance in utilitarian values was explained by Big Five personality (R^2= 0.09, R^2 (Adj) = 0.083, $F_{(245)}$ = 12.10, P<0.001) and openness (β= 0.146, P<0.05) played an effective role in the prediction of utilitarian values.

Table 2- Summaries of multiple regression analysis for Big Five personality predicting shopping motivations.

Dependent Variable	R^2	F	Df1	Df2	Significant Predictor(s)	β	Sig
Compulsive Buying	.027	6.777**	1	245	Conscientiousness	-.164**	.009
Impulsive Buying	.019	4.621*	1	245	Neuroticism	.136*	.033
Hedonic Values	-	-	-		-	-	-
Utilitarian Value	.09	12.104***	2	244	Conscientiousness	.223***	.000
					Openness	.146*	.023

* P<0.05; ** P<0.01; *** P<0.001

Canonical Correlation

Canonical correlation was performed which was one of the most appropriate tools when a researcher desire to examine the relationship between two variable sets (Sherry & Henson, 2005). Because canonical correlation is a multivariate technique, we could have more than one variable, so we put Big Five personality traits as predictor (independent) and shopping motivation variables as criterion (dependent). Table 3 shows the result of canonical correlation. Totally, the full model was statistically significant, with a Wilks's λ=0.81 ($F_{(20,\ 79031)}$ =2.52, P<0.001).. It is noticeable that Wilks's λ represents the amount of variance not explained by variable sets. Therefore, by taking 1- λ, we found the full model effect size in a r^2 metric (Joshanloo et al., 2012). Canonical correlation analysis output demonstrates four functions with canonical coefficients of 0.33, 0.24, 0.16 and 0.05 for each function. Function 1 was statistically significant ($F_{(20,79031)}$ =2.53, P<0.001). Although function 2 ($F_{(12,632,63)}$ = 1.76, P<0.051), ($F_{(6,480)}$ = 1.13, P<0.341), ($F_{(2,241)}$ = 0.32, P<0.729) didn't explain a significant amount of variance which shared between variable sets, we decided to consider function 2 in our analysis. The reason for this decision was rooted in Sherry and Henson's (2005) instruction for choosing functions. They suggested that researcher should only interpret those functions that explained reasonable amount of variable sets, and they advised researchers not to rely on statistical significance to decide whether the function should be interpreted (Function 1 explains about 11.18% of shared variance, function 2: 5.72%; function 3: 2.52% and function 4: 0.26). Table 3 demonstrates the structure coefficients are greater than |.3| from which, the standardized canonical coefficient, the communality coefficient, and the squared structured coefficient for function 1 and 2 can be interpreted. Structure coefficient of function 1 indicated that compulsive buying (r_s=-0.467), impulsive buying (r_s=-0.462) and utilitarian shopping values [illegible] 0.887) were interpretable contributors to the synthetic criterion variable. Considering the Big Five personality variable set as predicto[illegible] column in function 1 informs us that extraversion (r_s=0.427), agreeableness (r_s=-0.380), conscientiousness (r_s=0.935), neuroticism (r_s=-0.577) and openness (r_s=0.527) were contributors to the synthetic predictor variable. By analyzing the output of first canonical variable, we concluded that higher extraversion, agreeableness, conscientiousness and openness were associated with higher utilitarian values, but because of the negative structure coefficient for neuroticism, it was positively related to compulsive and impulsive buying and negatively related to utilitarian values. Moving on to function 2, the coefficient in Table 3 suggested the predictor variables of relevance were extraversion (r_s=-0.350), neuroticism (r_s=-0.426) and openness (r_s=0.465). On the other side, the only criterion variable of relevance was hedonic values (r_s=-0.617).

Table 3: Canonical solution for Big Five Personality predicting Shopping motivations for Functions 1 and 2.

	Function1			Function2			
Variable	**Coef**	r_s	r_S^2 (%)	**Coef**	r_s	r_S^2 (%)	h^2(%)
Predictors							
Extraversion	.024	.427	18.23	-.751	-.350	12.25	30.48
Agreeableness	-.156	.380	14.44	.263	.194	3.76	18.20
Conscientiousness	.830	.935	87.42	-.518	-.216	4.67	92.09
Neuroticism	-.266	-.577	32.29	-.600	-.426	18.15	50.44
Openness	.223	.527	27.77	.682	.465	21.62	49.39
R_c			33.44			23.92	
R_C^2			11.18			5.72	
Criterion							
Compulsive Buying	-.457	-.467	21.81	1.131	.081	.006	21.816
Impulsive Buying	-.100	-.462	21.34	-.291	-.196	3.84	25.18
Hedonic Values	.120	-.010	0.0001	-1.302	-.617	38.07	38.0701
Utilitarian Values	.836	.887	78.68	.431	.110	1.21	79.89

Note: Structure coefficients (r_s) greater than |.3| underlined. Coef = standardized canonical function coefficient; r_s = structure coefficient; r_s^2= squared structure coefficient; r_c = canonical correlation coefficient; r_c^2= squared canonical correlation coefficient; h^2= Communality coefficient

Second canonical variable indicated that higher extraversion and neuroticism were associated with higher hedonic values but, negatively related to openness. The last column listed the communality coefficient (h_2), which represented the amount of variance that was reproducible across the functions (Sherry et al., 2005). Communalities above 30% were also underlined to show the variable with the highest level of usefulness in the model. Considering Table 3 H_{a1} stated that a negative relationship between extraversion and impulsive buying was confirmed. In addition, H_{a2} was rejected which showed there was a negative relationship between extraversion and compulsive buying. The results confirmed H_{a3}, but rejected H_{a4} that showed a positive relationship between extraversion and utilitarian shopping values. About neuroticism and its relationship with shopping variables, H_{b1} hypothesized that the direct relationship between impulsive buying and neurotics was rejected and stated that this relationship was negative. In the case of other hypothesis, H_{b2} was rejected but H_{b3} and H_{b4}were confirmed. About the relationship between openness and shopping vaiables, despite the recent researches (Matzler, et al., 2006, Voss, et al., 2003), all of the hypotheses including H_{3a}, H_{3b}, H_{3c}, H_{3d} were rejected. It showed agreeableness had a negative relationship with compulsive and impulsive buying, positive relationship with hedonic shopping value and no significant relationship with utilitarian shopping values. Hypotheses H_{4a} H_{4b}, H_{4d}, which explored the relationship between agreeableness, compulsive buying, impulsive buying and utilitarian shopping values were confirmed but we didn't find a significant relationship between hedonic shopping values and agreeableness (H_{4c}). Analyzing Conscientiousness and shopping variables demonstrate that H_{5a}, H_{5b}, H_{5d} were confirmed, but no significant relationship between Conscientiousness and hedonic shopping values (H_{5c}) was found.

Supplemental Analysis

Sex Differences and Moderating Effect of Sex

In this part, we investigated the gender differences in personality and motivation. An independent-Sample t-Test was conducted to investigate gender differences. The results of t-Test indicated that males were more conscientiousness ($t_{(245)}$=2.51, P<0.05) and openness ($t_{(245)}$=2.03, P<0.05). On the other hand, females tended to be more neurotic ($t_{(245)}$= 2.634, P<0.01), they were more compulsive buyer ($t_{(245)}$= -3.671, P<0.001) and they did more shopping for hedonic purposes ($t_{(245)}$= -3.542, P<0.001). A hierarchical regression analysis was conducted to examine the impact of sex on the relationship between Big Five personality traits and shopping motivation variables. The results of the moderating effect of sex at step 3 showed that there was no significant interaction term as a moderator. It indicated that sex didn't have a moderating effect between personality traits and shopping motivation variables.

Sex Differences

In this part, we conducted the independent sample t-test to investigate the differences between males and females in order to make our conclusion more applicable and comprehensive. Table 4 showed the results that was based on the findings, males were more openness (t_{245}=-2.03 P<0.05). In contrast, females were more neurotic (t_{245}=-2.634 P<0.01), and more than males tended toward compulsive buying (t_{245}=-2.671 P<0.001) In the shopping process, females enjoyed much more than males and showed higher degree of hedonism (t_{245}=-3.542, P<0.001)

Table 4: Sex differences in personality traits and shopping

Variable	Independent sample t-test for analyzing means			Means		
	t	df	Sig	Males	Females	Mean Differences
Extraversion	-0.228	245	0.82	3.21	3.23	-0.017
Agreeableness	1.229	245	0.22	3.74	3.67	0.079
Conscientiousness	2.51*	245	0.013	3.54	3.36	0.187
Neuroticism	-2.63*	245	0.009	2.81	3.02	-0.210
Openness	2.038	245	0.043	3.63	3.49	0.141
Compulsive Buying-	3.67*	245	0.000	2.63	3.02	-0.390
Impulsive Buying	-1.397	245	0.164	2.34	2.46	-0.120
Hedonic Values	-3.542*	245	0.000	2.93	3.27	-0.343
Utilitarian Values	1.503	245	0.134	3.55	3.44	0.108

*P<0.05; **P<0.01; ***P<0.001

DISCUSSION

In this study, the relationship between Big Five Personality traits and shopping motivations was examined. Some parts of results supported the past research findings and some parts of them didn't support the recent research conclusions. Our findings showed that individual personality had an important role in the shopping procedure, but we had much more important factors in this procedure such as economic conditions, religion, social situation, governments' policies, cultural issues which directly or indirectly had an impact on forming individual's personality and shopping behaviors.

To better analyze the relationship between variables and control overlapping variance among them, we performed canonical correlation and regression analysis. Regression analysis output demonstrated that conscientiousness significantly predicted compulsive buying as a negative predictor. Individuals who got the high score on conscientiousness were able to control impulse emotions which were the background for the occurrence of compulsive behavior. Also, findings introduced neuroticism as a significant predictor of impulse buying. This was expectable that neurotics tended toward impulsiveness, because they were seeking for temporary ways to overcome distress. As mentioned, none of the personality traits significantly predicted hedonic shopping values. Furthermore, conscientiousness and openness were determined as the significant predictors of utilitarian shopping values. It wasn't strange to put conscientiousness as the predictor of utilitarian shopping values. But, introducing openness as a predictor of utilitarian was the opposite of the provided definition of openness provided by Costa & McCrae (1992c).

Canonical correlation analysis presented two significant and sufficiently explained variance functions. Function 2 indicated a paradox in convergence among neuroticism and extraversion. We mentioned that these two traits were almost opposite. Actually, the specification of neuroticism and extraversion is like the two ends of a spectrum. This paradox stated that both neurotic and extravert had tendency to shopping based on hedonism and their behavior in this shopping motivation was the same. In addition, this indicated that hedonic shopping was an irrelevant issue on social confidence level of individuals, because the differences between extraverts and neurotics in social confidence are remarkable (John et al., 2008; Pervin, 2006).

Nevertheless, our canonical correlation analysis outcome suggested that extraverts in Iran had a significant positive relationship with hedonic values. Besides, it reported a positive relationship between extraversion and utilitarian values, thus, like the conclusion of McCrae & Costa (2003), extraverts paid more attention to the benefits and fun-loving, passionate, active and positive emotion. Our findings indicated that there was a negative relationship between compulsive buying and extraversion. Although, DeSarbo and Edwards (1996) discussed about individuals who were in social isolation tended to compulsive buying in efforts to relieve the feelings of loneliness due to a lack of interaction with others, but Mowen (1999) found that there was no relationship between compulsive buying and extraversion and our findings supported the negative relationship result. Meanwhile, current research confirmed Zurawiki's (2010) description that the extraversion was inversely related to the thickness of the right anterior prefrontal cortex (PFC) and the right fusiform gyrus – regions possibly involved in the regulation of impulsive behavior.

Moreover, because conscientious and agreeable people are goal directed and they plan before doing any action, current research reported positive relationship between conscientiousness and agreeable individuals with utilitarian values, negative relationship with compulsive and impulsive buying and no significant relationship reported for hedonic values. In addition, an important implication of this study was rooted in neuroticism results. Our findings approved previous studies results such as Johnson et al., (2009); Mowen et al., (1999); d'Astous et al., (1990); Mick (1996) and Chetthamrongchai et al., (2000) that concluded a positive relationship between neuroticism with impulsive and compulsive buying and hedonic shopping values. The noticeable point was in female shopping behavior; the results showed females were more neurotic, impulsive, compulsive and hedonic shoppers than males. Another noticeable finding was that openness was inversely related to impulsive and compulsive buyings in the Iranian sample. This was inconsistent with theoretical definitions of openness provided by Pervin (2006) which defined them as the curiosity and the appreciation of experience for its own sake. This definition leads us to consider openness people as impulsive and compulsive buyers, but we confirmed Matzler (2006) hypothesis that higher openness individuals paid more attention to hedonic aspects of products.

In addition, we examined the moderated role of sex among shopping motivations and Big Five personality traits. Results indicated that sex didn't have a moderated role in this research. This means that sex wasn't an effective factor in shopping procedure among different personality traits.

Finally, we conducted an independence t-Test to determine the differences of personality traits and shopping motivations among males and females. The results were prominent and supported the prior researches. It showed males got higher scores on conscientiousness and openness than females. Besides, as mentioned, females got higher scores on neuroticism, compulsive buying and hedonic shopping values. Analyzed results confirmed the past researches which males tended to be more assertive and risk taking than females whereas females were generally higher than males in anxiety and tender-mindedness (Schmitt et al., 2008). As mentioned, it stated that females tended to be more neurotic and this tendency emerged in compulsive buying. In addition, we mentioned that neurotics were more hedonic, so females are tended to be more hedonic shoppers. Prior researches concluded that males were risk taking and tended to explore unfamiliarity. Our findings supported these results and showed that males tended to be more open toward discovering new areas.

The present study confronted some limitations. One of them was the sample we used. For further studies, we suggest to use different demographics and different university degrees. This can be useful to investigate education's role in personality traits and shopping motivations. Another limitation was about the method of research. Unfortunately, our participants just answered the questions without the perception of conditions. It will be useful and more precious to conduct an empirical investigation by applying experimental and quasi-experimental methods. In addition, we noticed that some items of scales weren't suitable for Iranian society and culture, especially (Persian) translated version of Big Five personality scale wasn't a suitable tool to evaluate personality in Iran. Prior researches about the translated version of Big Five personality approved this claim (Haghshenas, 1999; Joshanlooet al., 2010). Unfortunately, the scale reliability of utilitarian shopping value didn't report sufficient amount of reliability to get minimum levels of confirmation (Churchill Jr., 1979). Despite these limitations, this study found associations among Big Five personality traits and shopping motivations whereas this association wasn't moderate. Also, gender difference analysis emphasized on prior research findings.

However, this study had a prominent contribution to the interdisciplinary area of marketing and psychology. Although several researches had been conducted to investigate the relationship between personality traits and shopping motivations, the current study employed whole shopping motivation variables which some of which were used in prior research. This allowed us to hold more comprehensive investigations toward further consumer behavior and personology studies.

MANAGERIAL IMPLICATIONS

In this study, we investigated the impact of personality traits on shopping behaviors and motivations. We conducted t-test to examine the differences between males and females. The results helped managers to set goals and to choose right customers based on their products and personality. For example, the results demonstrated that females were more neurotic and impulsive buyers, but they paid much more attention to the family economic issues (Sweeney, 2002), so manufacturers should produce cheap goods with acceptable quality. Males were more utilitarian and showed more openness (3.55>3.44). They needed goods and services which were more creative, innovative and considered its price, so that products should have more equipment and in some cases, should be multipurpose. Furthermore, we can use personality traits in computer and console game industry. For example, Japan, Korea, China and South-East Asian countries tend to do their task in work teams, and they have high scores on agreeableness (Robbins & Judge, 2014). But Western countries pay much more attention to individualism, so game studios can produce team work games and PRG (Rol Playing Games) for each specific region. Some businesses such as shopping malls can provide very cheap but good quality products to satisfy their customer's impulsive needs when they are neurotic. Also, the results showed that impulsive buying could drive people to compulsive buyers, if it continued.

REFERENCES

1. Babin, B. J., Darden, W. R., & Griffin, M. (1994). Work and/or Fun: Measuring Hedonic and Utilitarian Shopping Value. Journal of Consumer Research, 20(4), 644-656.
2. Balabanis, G. (2002). The relationship between lottery ticket and scratch-card buying behaviour, personality and other compulsive behaviours. Journal of Consumer Behaviour, 2(1), 7-22.
3. Baumgartner, H. (2002). Toward a Personology of the Consumer. Journal of Consumer Research, 29(2), 286-292.
4. Buss, D. M. (2008). Human Nature and Individual Differences: Evolution of Human Personality. In O. P. John, R. W. Robins, & L. A. Pervin (Eds.), Handbook of Personality: Theory and Research (Third ed., pp. 29-56). New York: The Guilford Press.
5. Carpenter, J. M., & Moore, M. (2009). Utilitarian and hedonic shopping value in the US discount sector. Journal of Retailing and Consumer Services, 16(1), 68–74.

6. Chetthamrongchai, P., & Davies, G. (2000). Segmenting the Market for Food Shoppers Using Attitudes to Shopping and to Time. British Food Journal, 101(2), 81-101.
7. Churchill Jr., G. A. (1979). A Paradigm of Developing Better Measures of Marketing Construct. Journal of Marketing Research, 16(1), 64-73.
8. Costa, Jr, P. T., & McCrae, R. R. (1992b). Normal Personality Assessment in Clinical Practice: The NEO Personality Inventory. Psychological Assessment, 4(1), 5-13.
9. Costa, Jr., P. T., & McCrae, R. R. (1995). Domains and Facets: Hierarchical Personality Assessment Using the Revised NEO Personality Inventory. Journal of Personality Assessment, 64(1), 21-50.
10. Costa, P. T., & McCrae, R. R. (1992a). Revised NEO Personality Inventory (NEO-PI-R) and NEO Five-Factor Inventory (NEO-FFI) professional manual. Odessa, FL: Psychological Assessment.
11. Costa, P. T., & Robert, R. M. (1985). The NEO Personality Inventory manual.
12. Odessa, FL: Psychological Assessment.
13. Cramer, D. (2003). Advanced Quantitative Data Analysis. Philadelphia: Open University Press.
14. d'Astous, A., Maltais, J., & Roberge, C. (1990). Compulsive Buying Tendencies of Adolescent Consumers. Advances in Consumer Research,
15. 17, 306-312.
16. Dattalo, P. (2008). Determining Sample Size: Balancing Power, Precision, and Practicality. New York: Oxford University Press.
17. Davey, A., & Savla, J. (2010). Statistical power analysis with missing data: a structural equation modeling Approach. New York: Taylor & Francis.
18. DeSarbo, W. S., & Edwards, E. A. (1996). Typologies of Compulsive Buying Behavior: A Constrained Clusterwise Regression Approach. Journal of Consumer Pychology, 5(3), 231-262.
19. DeYoung, C. G., Peterson, J. B., & Higgins, D. M. (2005). Sources of Openness/Intellect: Cognitive and Neuropsychological Correlates

of the Fifth Factor of Personality. Journal of Personality, 73(4), 825-858.

20. Digman, J. M. (2002). Historical Antecedents of the Five-Factor ModeL. In P.
21. T. Costa, & T. A. Widiger, Personality Disorders and the Five-Factor Model of Personality (pp. 3-7). American Psychological Association.
22. Duijsens, I. J., & Diekstra, R. F. (1996). DSM-III-R and ICD-10 personality disorders and their relationship with the big five dimensions of personality. Personality and Individual Differences, 21(1), 119-133.
23. Eren, S. S., Eroglu, F., & Hacioglu, G. (2012). Compulsive buying tendencies through materialistic and hedonic values among college students in Turkey. Procedia - Social and Behavioral Sciences, 58, 1370 – 1377.
24. Faber, R. J., & O'Guinn, T. C. (1989). Classifying Compulsive Consumers: Advances in the Development of a Diagnostic Tool. Advances in Consumer Research, 16, 738-744.
25. Faber, R. J., & O'Guinn, T. C. (1992). A Clinical Screener for Compulsive Buying. Journal of Consumer Research, 19(3), 459-469.
26. Faber, R. J., O'Guinn, T. C., & Krych, R. (1987). Compulsive Consumption. Advances in Consumer Research, 14, 132-135.
27. Faul, F., Erdfelder, E., Buchner, A., & Lang, A. G. (2009). Statistical power analyses using G*Power 3.1: Tests for correlation and regression analyses. Behavior Research Methods, 41(4), 1149-1160.
28. Goldberg, L. R. (1990). An Alternative "Description of Personality": The Big-Five Factor Structure. Journal of Personality and Social Psychology, 59(6), 1216-1229.
29. Guido, G., Capestro, M., & Peluso, A. M. (2007). Experimental Analysis of Consumer Stimulation and Motivational States in Shopping Experiences. International Journal of Market Research, 49(3), 365-386.
30. Gustavsson, J. P., Jonsson, E. G., Linderb, J., & Weinryb, R. M. (2003). The HP5 inventory: definition and assessment of five health

relevant personality traits from a five-factor model perspective. Personality and Individual Differences, 35(1), 69–89.

31. Haghshenas, H. (1999). Standardization of NEO PI Revised. Andisheh va Raftar, 4(4), 38-47 (In Persian).
32. Hausman, A. (2000). Multi-method investigation of consumer motivations in impulse buying behavior. Journal of Consumer Marketing, 17(5), 403 - 426.
33. Hirschman, E. C., & Holbrook, M. B. (1982). Hedonic Consumption: Emerging Concepts, Methods and Propositions. Journal of Marketing, 46(3), 92-101.
34. Ho, V. T., Weingart, L. R., & Rousseau, D. M. (2004). Responses to broken promises: Does personality matter? Journal of Vocational Behavior, 65(2), 276–293.
35. John, O. P., Donahue, E. M., & Kentle, R. L. (1991). The Big Five Inventory-Versions 4a and 54. Berkeley, CA: University of California,Berkeley, Institute of Personality and Social Research.
36. John, O. P., Naumann, L. P., & Soto, C. J. (2008). Paradigm Shift to the Integrative Big Five Trait Taxonomy: History, Measurement, and Conceptual Issues. In O. P. John, R. W. Robins, & L. A. Pervin (Eds.), Handbook of Personality: Theory and Research (3rd ed., pp. 114-158). New York, NY: Guilford Press.
37. Johnson, T., & Attmann, J. (2009). Compulsive buying in a product specific context: clothing. Journal of Fashion Marketing and managent, 13(3), 394405.
38. Joshanloo, M., Daemi, F., Bakhshi, A., Nazemi, S., & Ghafari, Z. (2010).
39. Construct Validity of NEO-Personality Inventory-Revised in Iran. Iranian
40. Journal of Psychiatry and Clinical Psychology, 16(3), 220-230 (In Persion).
41. Joshanloo, M., Rastegar, P., & Bakhshi, A. (2012). The Big Five personality domains as predictors of social wellbeing in Iranian university students. Journal of Social and Personal Relationships, 29(5), 639–660.
42. Kacen, J. J., & Lee, J. A. (2002). The Influence of Culture on

Consumer

43. Impulsive Buying Behavior. Journal of Consumer Psychology, 12(2), 163176.
44. Kim, B., & Han, I. (2011). The role of utilitarian and hedonic values and their antecedents in a mobile data service environment. Expert Systems with Applications, 3, 2311–2318.
45. Lakhal, S., Frenette, É., Sévigny, S., & Khechine, H. (2012). Relationship between choice of a business major type (thing-oriented versus personoriented) and Big Five personality traits. The International Journal of Management Education, 10(2), 88–100.
46. List of countries by income equality. (n.d.). Retrieved from Wikipedia: http://en.wikipedia.org/wiki/List_of_countries_by_income_equality
47. Maltby, J., Day, L., & Macaskill, A. (2010). Persionality, Individual Differences and Intelligence. Essex: Pearson.
48. Matzler, K., Bidmon, S., & Grabner-Kräuter, S. (2006). Individual determinants of brand affect: the role of the personality traits of extraversion and openness to experience. Journal of Product & Brand Management, 15(7), 427-434.
49. McCrae, R. R., & Costa, Jr, P. T. (2003). Personality in Adulthood; A FiveFactor Theory Perspective (2nd ed.). New York: Guilford Press.
50. Mick, D. G. (1996). Are Studies of Dark Side Variables Confounded by Socially Desirable Responding? The Case of Materialism. Journal of Consumer Research, 23(2), 106-119.
51. Mowen, J. C., & Spears, N. (1999). Understanding Compulsive Buying Among College Students: A Hierarchical Approach. Journal of Consumer Psychology, 8(4), 407-430.
52. Nguyen, T. T., Nguyen, T. D., & Barrett, N. J. (2007). Hedonic shopping motivations, supermarket attributes, and shopper loyalty in transitional markets: Evidence from Vietnam. Asia Pacific Journal of Marketing and Logistics, 19(3), 227 - 239.
53. Pervin, L. A. (2006). The Science of Personality (3rd ed.). New York: Oxford University Press.

54. Rintamäki, T., Kanto, A., Kuusela, H., & Spence, M. T. (2006). Decomposing the value of department store shopping into utilitarian, hedonic and social dimensions: Evidence from Finland. International Journal of Retail & Distribution Management, 34(1), 4-26.
55. Robbins, S. P., & Judge, T. A. (2014). Organizational Behavior (16 ed.). NewYork: Prentice Hall.
56. Roberts, J. A., & Pirog, S. F. (2004). Personal Goals and Their Role in
57. Consumer Behavior: The Case of Compulsive Buying. Journal of Marketing Theory and Practice, 12(3), 61-73.
58. Rook, D. W. (1987). The Buying Impulse. The Journal of Consumer Research, 14(2), 189-199.
59. Rook, D. W., & Fisher, R. J. (1995). Normative Influences on Impulsive Buying Behavior. Journal of Consumer Research, 22, 305-313.
60. Schmitt, D. P., Realo, A., Voracek, M., & Allik, J. (2008). Why Can't a Man Be More Like a Woman? Sex Differences in Big Five Personality Traits across 55 Cultures. Journal of Personality and Social Psychology, 94(1), 168–182.
61. Sherry, A., & Henson, R. K. (2005). Conducting and Interpreting Canonical Correlation Analysis in Personality Research: A User-Friendly Primer. Journal of Personality Assessment, 84(1), 37-48.
62. Sweeney, M. M. (2002). Two Decades of Family Change: The Shifting Economic Foundations of Marriage. American Sociological Review, 67(1), 132-147.
63. Teimourpour, B & ,.Hanzaee, K. H .(2014) .An analysis of Muslims 'luxury market in Iran: Challenges and opportunities .Journal of Islamic Marketing.209-198 ,(2)5 ,
64. Verplanken, B., & Herabadi, A. (2001). Individual Differences in Impulse Buying Tendency: Feeling and no Thinking. European Journal of
65. Personality, 15(1), 71-83.

66. Voss, K. E., Spangenberg, E. R., & Grohmann, B. (2003). Measuring the Hedonic and Utilitarian Dimensions of Consumer Attitude. Journal of Marketing Research, 40(3), 310-320.
67. Wang, C.-C., & Yang, H.-W. (2008). Passion for Online Shopping: the Influence of Personality and Compulsive Buying. Social Behavior and Personality, 36(5), 693-706.
68. World Economic Outlook Database April 2014. (n.d.). Retrieved from IMF:http://www.imf.org/external/pubs/ft/weo/2014/01/weodata/weorept.as px
69. Zurawicki, L. (2010). Neural Bases for Segmentation and Positioning. In Neuromarketing (pp. 163-178). New York: Springer.

CHAPTER 8

THE RELATIONSHIP BETWEEN GROWTH-INEQUALITY-POVERTY TRIANGLE AND ENVIRONMENTAL DEGRADATION: UNVEILING THE REALITY

Syeda Anam Hassan, Khalid Zaman and Shah Gul

Department of Management Sciences, COMSATS Institute of Information Technology, Abbottabad, Pakistan.

ABSTRACT

It is important to recognize that increase in well-being is no longer dependent on further economic growth, but on economic and social well-being, which means that the decline in carbon dioxide emission without reducing real wellbeing and growth. The aim of this study is to investigate the main driving forces affecting short and long-run carbon emissions pattern due to changes in growth, inequality and poverty triangle in Pakistan over the period 1980 – 2011 by using multivariate cointegration approach. This study uses five different models i.e., each model have an important policy implication in the context of Pakistan. The results indicate that, on the short run, there is a significant negative relationship between economic growth & carbon emissions and economic growth & poverty while there is a positive relationship between i) economic growth & income inequality; and ii) poverty & income inequality. On the long-run, there is a significant positive relationship between GDP & income inequality, carbon dioxide emissions & income inequality, and poverty & income inequality in Pakistan. On the other side, there is a negative relationship between carbon emissions & economic growth, carbon emissions & income inequality, and economic growth & income inequality. The results of environmental Kuznets curve (EKC) hypothesis show an inverted U-shaped trajectory in relation to economic growth in Pakistan. This study contributes to the debate on the existence and policy relevance of the EKC for Pakistan. The conclusion ensures the sustainability of an urgent need to look beyond the EKC by adopting courageous policy measures of environmental preservation in Pakistan irrespective of the country's level of income. For reduction of CO2 emission, environmental progressive management policies, economic transport system regulations, and low emit fuel consumption by industries are the need of the world.

INTRODUCTION

Global environmental degradation has been inspiring more and more researchers to investigate the causes of environmental degradation. There is a complex relationship between environmental changes and its driving forces including economic growth and environmental degradation (McPherson and Nieswiadomy, 2005).

Environmental degradation is a process through which the natural environment is compromised in some way, reducing biological diversity and the general health of the environment. This process can be entirely natural in origin, or it can be accelerated by human activities. Many international organizations recognize environmental degradation as one of the major threats facing the planet, since humans have only been given one Earth to work with, and if the environment becomes irreparably compromised, it could mean the end of human existence (Smith, 2013).

The natural resources are consumed and waste is produced at sustainable rates represent major pertaining challenges. Recognition of these challenges resulted in the endorsement in 2000 of environmental sustainability as one of the Millennium Development Goals (MDGs) to be achieved by 2015. However, by 2003 global rates of consumption and waste production were estimated to be at least 25% higher than the capacity of the planet to provide resources and absorb waste and this rate may have risen as high as 50% by 2007 (WWF, 2010).

On another perspective when the environment becomes polluted, it means that toxic substances have rendered it unhealthy. Pollution can come from a variety of sources, including vehicle emissions, agricultural runoff, accidental chemical release from factories, and poorly-managed harvesting of natural resources. In some cases, pollution may be reversible with costly environmental remediation measures, and in other instances, it may take decades or even centuries for the environment to cope with the pollution (Wallace, 2013).

Since the Industrial Revolution, human activities such as the burning of oil, coal and gas, as well as deforestation have greatly increased CO_2 concentrations in the atmosphere. Almost all CO_2 emissions (about 96.5%) come from fossil fuels use. The three types of fossil fuels that are used the most are coal, natural gas and petroleum (oil). When fossil fuels are combusted, the carbon stored in them is emitted almost entirely as CO_2 (U.S Department of Energy, 2007).

The most important source of CO_2 emissions worldwide is caused by the transportation of goods and people. The emissions caused by people traveling (by car, plane, train, etc.) are examples of direct emissions. The emissions caused by the transportation of goods are examples of indirect emissions since the consumer has no direct control of the distance between the factory and the store. Since the distance between the manufacturer and the consumer is constantly growing, more pressure is put on the transportation industry to bridge this gap and this ends up creating more indirect emissions. What's worse is that 99% of the energy used to transport people and goods all over the world comes from the combustion of fossil fuels (UDE, 2007). All industrialized nations (with the exception of Canada and France) get the majority (between 60-80%) of their electricity from the combustion of fossil fuels. Eighty percent of the world's energy comes from fossil fuels. The current burning of fossil fuels – oil/ petroleum, coal and natural gas - releases 7 billion tons of carbon per year in the form of carbon dioxide, plus lots of other greenhouse gases (EPA, 2008).

Manufacturing and industrial processes all combine to produce large amounts of each type of greenhouse gas but specifically large amounts of CO_2 because of two reasons. Firstly, many manufacturing facilities directly use fossil fuels to create heat and steam which is needed at various stages of production. Second, their energy intensive activities use more electricity than any other sector so unless they are using renewable sources the energy that they use is responsible for vast amounts of emissions. By industrial production means manufacturing, construction, mining, and agriculture. Manufacturing is the largest of the four and can be broken down into five main categories: paper, food, petroleum refineries, chemicals, and metal/mineral products. These categories account for the vast majority of the energy use and CO2 emissions by the sector (AER, 2004).

CO2 when measured in absolute term, by nation-state, then China ranks first and is outpacing the rest of the world fast, with the United States second, and then Indonesia and Brazil, with a few others coming behind. If rankings are based on per-capita emissions, then Western European stateliest and America lead the pack, and far, far, far behind them, the large developing states of the global South (Ajl, 2010).

Generally at global level, the huge power stations generating electricity to keep lights on, industry operating. They're burning coal, natural gas and oil. They account for roughly one-third of carbon dioxide production in the industrialized nations. Secondly, CO_2 emissions from transport sector accounts mostly about 20 to 25 percent in most nations. It includes cars, buses, planes, trains, trucks. Behind transport and power generation the home heating, agriculture, industries like the cement industry, which is a very polluting industry because it has to heat up limestone to 1,450 degrees Celsius to turn it into cement, and the chemical process itself produces carbon dioxide. Cement is 5 to 10 percent of global CO_2 emissions (Monbiot, 2008).

Carbon dioxide (CO_2) emission in the United States is increased by about 10% between 1990 and 2011. Since the combustion of fossil fuel is the largest source of greenhouse gas emissions in the United States, changes in emissions from fossil fuel combustion have historically been the dominant factor affecting total U.S. emission trends. Changes in CO_2 emissions from fossil fuel combustion are influenced by many long-term and short-term factors, including population growth, economic growth, changing energy prices, new technologies, changing behaviour, and seasonal temperatures. Between 1990 and 2011, the increase in CO2 emissions corresponded with increased energy use by an expanding economy and population, and an overall growth in emissions from electricity generation. Transportation emissions also contributed to the 10% increase, largely due to an increase in miles travelled by motor vehicles (US department, 2007).

In 1987, the World Commission on Environment and Development (WCED) suggested "Poverty is a major cause and effect of global environmental problems. It is therefore futile to attempt to deal with environmental problems without a broader perspective that encompasses the factors underlying world poverty and international inequality." To better coordinate the process of economic and social development and environmental conservation, it is vital to understand the link between the two. They can't be viewed as two independent phenomena, as there assert a cause and effect relationship between them. Several regions and communities of the world are trapped in this vicious spiral where poverty pushes people to over exploit the environmental resources to survive each day. The impoverishment of the ecosystem further pushes them deeper into poverty making survival an even more difficult task. (Jabeen, 2012).

Environmental degradation appears to be another side effect of economic inequality and analyses show there is a negative correlation between income inequality and environmental sustainability the higher the income inequality the worse the environmental indicators such as waste production, meat and water consumption, biodiversity loss and environmental composite indices (e.g. ecological footprint) (Andrich et al., 2010).

It is well known that higher level of pollution emission (carbon dioxide emission) might lead to the reduction of the productive capacity of a country and also to climatic change. On the other hand it is also a fact economic growth necessitate higher amount of energy consumption and thus carbon dioxide and other pollution emission. In fact it has been argued that an inverted-U-shaped relationship between economic growth and measured pollution indicators (environmental quality) exist and this known as the EKC. The Environmental degradation and Economic developmental stages are best described in relationship by an inverted U-shape curve (the Environmental Kuznets Curve (EKC)). From this perspective, as the economy expands, environmental damage increases up to a threshold, above which the society demands and can afford environmental remediation and natural resource protection. In industrial period the environmental damages are at peak and after then it declines. Therefore Environmental degradation and economic development stages follow the pattern of Inverted U shape phenomena such that increase peak and decline rate with economic growth. Finally, if the environmental Kuznets curve hypothesis is supported by evidence, development policies have the potential of being environmentally benign over the long run (at high incomes), but they are also capable of significant environmental damage (Papanyotou, 1994).

Boyce et al (2007) have proposed that the unequal distribution of wealth and power within countries leads to greater environmental damage by undermining the collective action required for environmental protection. Likewise, it has been shown that more equal societies are more socially cohesive and have higher levels of trust which foster public-spiritedness (Wilkinson & Pickett, 2010). Similar findings have been reported on the relationship between social capital and environmental outcomes (Dulal et al., 2011), presumably because lack of trust and poorer social capital erode the capacity for collaborative action.

The annual carbon emissions and per capita emissions for Pakistan are expressed in thousand metric tons of carbon. Pakistan produces more than 30 million metric tons of carbon emissions. This is about 0.4 per cent of global emissions. But, this has increased almost four-fold from nine million in 1980. A heavy emphasis on industrialization in Pakistan means that the rate of increase in emissions is going up. The energy sector contributes the most to emissions is 53 per cent. At the same time, Pakistan's decreasing forest cover is suffering from among the world's worst deforestation rates, primarily from the large logging industry. Forest cover declined to 2.5 per cent in 2005 from 3.3 per cent in the late 1990s (Sheikh, 2008).

Environmental challenges and issues of Pakistan are associated primarily with an imbalanced social and economic development from the last two to three decades. This challenge is further compounded with rapid urbanization due to a shift of population from rural to urban areas. Thus, all major cities of Pakistan face haphazard, unplanned expansion leading to increase in pollution. Main factors causing degradation to air quality are, a) rapidly growing energy demand and b) a fast growing transport sector. In the cities, widespread use of low quality fuel, combined with a dramatic expansion in the number of vehicles on roads, has led to significant air pollution problems. Air pollution levels in Pakistan's most populated cities are high and climbing causing serious health issues. Although Pakistan's energy consumption is still low by world standards, but lead and carbon emissions are major air pollutants in urban centers (GOP, 2010).

The above discussion shows the strong relationship between poverty-growth-inequality and environmental degradation. The objective of this study is:

- To observe changes in environmental degradation due to changes in poverty, growth and income inequality in Pakistan.
- To examine the short-and long-run relationship among carbon dioxide emissions; poverty, economic growth and income inequality.
- To find the Existence of environmental Kuznet Curve for the Pakistan.

This study divides in to the following sections: after introduction which is presented in Section 1 above, Section 2 presents the review of literature.

Data source and methodological framework are shown in Section 3. Results are discussed in Section 4. Final section concludes the study.

LITERATURE REVIEW

The relationship between CO_2 emission and growth-inequality-poverty has been the subject of intense research. Some earlier panel studies reveal different results which depend upon the countries and the period held in the analysis. As the links between economic growth and CO_2 emissions have been primary concerns of the empirical research, particularly since the 1990s. A number of empirical studies have enumerated the possible linkages of CO2 emissions and economic growth.

Ahmed (2001) attempt to study the income inequalities between and within various occupations/professions in Pakistan utilizing Household Income and Expenditure Survey (HIES) 1992-93. The results show that within various occupations/ professions, significant and highest level of inequality among skilled workers and lowest level of inequality is found among professionals. Similarly, within various professions/occupations, highest level of inequality is observed in NWFP while lowest level of inequality is found in Baluchistan Province.

Roco and Alcantara (2001) studied the relationship between CO_2 emissions and primary energy in case of Spain in time period 1972 to 1997. The results show that there is no significant evidence supporting the Kuznets curve hypothesis. Tabassum (2004) examines the relationship between economic growth and income inequality both at aggregate and regional level using more comparable data set for 69 developing countries over the period 1965-2003. The study identifies credit market imperfection in low-income developing countries as the likely reason for a strong negative relationship between income inequality and economic growth. While in short run the relationship between growth and income inequality is positive but over time more income inequalities reduces economic growth. Moreover, this study finds evidence that more physical and human capital investment, Openness to trade and higher government spending have statistically significant impact on enhancing economic growth and reducing inequality.

Khanam (2007) studied the relationship between primary education and poverty reduction which affects the individual income by using Mincer Wage function. The experiment results show direct effects of primary education on income poverty reduction at earning member levels, at household levels and at per capita levels. The income distributions in the two years have variable effects at income ranges however income distributions at different percentiles of population are affected positively by primary education expansion. The results further show that as primary education expands, the income levels increase within income quintiles and the income inequality reduces in 2001-2002 cases. It is also observed that higher the changes in population schooling levels, higher will be the changes in income inequalities and poverty profiles.

Acharyya (2009) examines two most important benefits and costs of foreign direct investment in the Indian context - GDP growth and the environment degradation during 1980 to 2003. The findings show there is statistically significant long run positive relation, but marginal, impact of FDI inflow on GDP growth. While the long run, growth impact of FDI inflow on CO_2 emissions is quite large. The finding has some far reaching implications for the global environment as well, with India having emerged as the fourth highest in the global ranking of CO2 emissions by turn of this century. The results from cointegration analysis show that FDI inflow in India did have a positive, but marginal, long run impact on GDP growth. On the other hand, though the pollution heaven hypothesis may not be a plausible argument for the upsurge in FDI inflow in the 1990s, such inflows did have a quite large positive impact on the CO2 emissions through output growth. This, however, may be an underestimate since the effect of FDI on water pollution and on the local air pollutants are not estimated.

Burnett (2009) investigates the relationship between economic growth and environmental indications for air pollution in over 100 metropolitan statistical areas from 2001 to 2005 for the United States. The results indicate a statistically significant U-shaped relationship for some of the pollutants; however, the evidence is pretty weak with the exception of ground level ozone. This study does not find evidence to support the traditional EKC inverse U-shaped relationship. Moreover these results are compared and contrasted to previous literature by providing insight into unresolved theoretical and empirical estimation issues and future research needs.

Shahbaz et al (2009) investigate the relationship between CO2 emissions, energy consumption, economic growth and trade openness in the context of Pakistan over a period of 1971 to 2009. By using ARDL model for cointegration and Granger causality tests the result infers that there exists long run relationship among the variables. The EKC hypothesis is supported in the country. Furthermore, the consequences find one-way causal relationship in succession from income to CO2 emissions. Energy consumption increases CO2 emissions both in short and long run. Trade openness reduces CO2 emissions in long run but it is insignificant in short run.

Boopen and Vinesh (2010) analyses the relationship between GDP and carbon dioxide emissions for Mauritius for the period 1975 to 2009. By Using Solow Growth function pertaining to the economic model and rigorous econometrics analysis, results depicts that the carbon dioxide emission trajectory is closely related to the GDP time path. The estimation shows that emissions elasticity on income has been increasing over time. By estimating the EKC, finding is unable to prove the existence of a reasonable turning point and thus no EKC "U" shape is obtained. Findings imply that Mauritius could not curb its carbon dioxide emissions in the last three decades. Thus, as hypothesized, the cost of degradation associated with GDP grows over time and it advocate that the economic and human activities are having increasingly negative environmental impacts on the country as compared to their economic prosperity.

Drabo (2010) examines the link between health indicators, environmental variables and income inequalities in 90 developed and developing countries by using data covering the period from 1970 to 2000. The econometric analysis shows that income inequalities negatively affect environmental quality and environment degradation worsens population's health. This negative effect of income inequalities on environment is mitigated by good institutions. The findings also show that income inequalities negatively affect health status. This confirms that environment quality is an important channel through which income inequalities affect population health. These results hold for air pollution indicators (CO2 and SO2) and water pollution indicator (BOD). The results are also robust for rich and developing countries. Countries with high income inequalities may implement distributive policies in order to avoid its negative impact on health.

Leight (2010) investigates the relationship between economic inequality, income growth, and political control in US states for the time frame 1969 to 2005. The result shows that in the short run, inequality has a significant and robust impact on growth that is robust to a number of different control variables and estimation techniques. In addition the findings reveal that the relationship among inequality is non-linear; a quadratic function relating the Gini coefficient to growth is often significant. When other measures of inequality are used in place of the Gini coefficient, they do not exhibit the same relationship with growth. Finally, political control appears to have a significant impact on both growth and inequality. Democratic control increases growth and reduces inequality, while Republican control has the opposite effects.

Murad and Mustapha (2010) examines whether or not the waste management practices of the poor households living in squatters and low-cost flats in Kuala Lumpur are conducive to the environment. The aim of this study is to empirically assess knowledge, attitude and behaviour of the urban poor concerning their household solid waste management. With primary data collected from the level of living condition and waste management practices of the urban poor, the findings of the study provide evidence to the effect that poverty does not cause environmental degradation as the knowledge, attitude, and behaviour of the urban poor concerning solid waste management are found to have been conducive to the environment.

Attari and Attaria (2011) investigate the effect of CO2 emission on economic growth by conducting the regional analysis of PIC nations i.e. Pakistan, India and China for the time period 1971 to 2008. The results show that the change in CO2 emission from 1971 to 2008 is at the rate of 51.33% globally. But in case of Pakistan it increased with minimal low rate which is 85.87% as compared to its neighbour countries like China and India which had 87.53% and 88.18% respectively. The population contribution by Pakistan is high, 63.55%, but GDP per capita is low. But the emission of CO2 from the consumption of energy is low as compared to China and India. Furthermore, the per capita emission of CO2in case of Pakistan is low as compared to China and India.

Essien (2011) examines the relationship between economic growth and CO2 emissions in Nigerian economy over the period 1980 to 2009. The econometric result suggests that there exists a long run relationship among the variables. Moreover the results reveals that electricity and gas consumption cause economic growth both in the short and long run but only fuel woods influences it in the long run. Finally the results provide confirmation that natural gas influences carbon emissions in the long run while fuel woods influences carbon emissions in the short run. These results are anticipated to shed a light on the Nigerian policy making in coping with climate change.

Mekenbayeva and Karakuş (2011) examine the relationship between Economic inequality and growth for 136 countries over a period of 1980 to 2009. A negative relationship of two variables in developing countries can result from negative opportunity-creation effect operating through the constraints on human capital investment that it imposes on poor individuals, imperfect capital markets, unsuitable and risky conditions for investors and unstable political environment. While the situation in develop countries is different because they have higher marginal productivity of capital, higher marginal propensity to save in general and a weaker preference for leisure. So far, Turkey is concerned the negative correlation between income growth and inequality due to development policies.

Ozturk and Acaravci, (2011) studied the CO2 emissions, energy consumption, and economic growth in Turkey for the period 1968 to 2005. The results suggest an evidence of a long-run relationship between the variables at 5% significance level in Turkey. The estimated income elasticity of carbon emissions per capita is -0.606 and the income elasticity of energy consumption per capita is 1.375. Results for the existence and direction of Granger causality show that neither carbon emissions per capita nor energy consumption per capita cause real GDP per capita, but employment ratio causes real GDP per capita in the short run. Moreover, EKC hypothesis at causal framework by using a linear logarithmic model is not valid in Turkish case. The overall results indicate that energy conservation policies, such as rationing energy consumption and controlling carbon dioxide emissions, are likely to have no adverse effect on the real output growth of Turkey.

Tiwari (2011) examined causality by considering energy consumption, C02 emissions and economic growth for India in period 1971 to 2005. The results show that CO2 Granger-causes GDP while energy consumption does not Granger- cause GDP, GDP does not Granger-cause CO2 while energy consumption Granger-causes CO2 emissions. The econometric results show that CO2 Granger-causes GDP while electricity consumption does not Granger-causes GDP in short run. In the long run, it is found that cointegrating vector of CO2 equation of VECM is significant. This implies that GDP, CO2, and electricity consumption Granger-cause CO2 emissions in the long run.

Zaman et al (2011) studied the relationship between poverty, population growth and environment. The aim of this study is to examine the impact of poverty on environment (air pollution) and population in context of Pakistan during a period of 1975 to 2009. The results show that rapid population and air pollution has a significant contributor to poverty in Pakistan. However, the results nullify the conventional view that poverty is a major cause of environmental degradation (or air pollution), while the result supports the hypothesis that population have a deleterious impact on increasing poverty. The results further conclude that there is a stable long-run relationship between population, poverty and pollution in Pakistan. The results of causality test show that there is a unidirectional causal flow from population to carbon dioxide emission.

Alkhathlan, et al (2012) examines the long run and causal relationship between economic growth, CO2 emissions and energy consumption in a multivariate framework for Saudi Arabia during the year 1980-2008. By using ARDL and VECM test the results show that there are long run and short run relationships among the variables in the model. The estimated elasticity coefficients of CO2 emissions, energy consumption, and employment ratio have positive and significant impacts on GDP in the long run. Additionally results for direction of causality indicate that neither carbon emissions per capita nor energy consumption per capita causes GDP per capita, but employment ratio causes GDP per capita in the short run. On the foundation of empirical findings, it can be accomplished that energy conservation policies and controlling carbon dioxide emissions, are likely to have no adverse effect on economic growth in the short run. Nevertheless, the long run income elasticity of carbon emissions is greater than the short run income elasticity of carbon emissions, which indicates that in the long run income leads to greater carbon dioxide emissions in the country.

Arouri et al, (2012) attempt to investigate the relationship between carbon dioxide emissions, energy consumption, and real GDP for 12 Middle East and North African Countries (MENA) over the period 1981 to 2005. The results show that in the long-run energy consumption has a positive significant impact on CO2 emissions. More interestingly, findings show that real GDP exhibits a quadratic relationship with CO2 emissions for the region as a whole. However the estimated long-run coefficients of income and its square satisfy the EKC hypothesis in most studied countries, the turning points are very low in some cases and very high in other cases, hence providing poor evidence in support of the EKC hypothesis. Thus, findings concludes that not all MENA countries need to sacrifice economic growth to decrease their emission levels as they may achieve CO2 emissions reduction via energy conservation without negative long-run effects on economic growth.

Cheema and Sial (2012) determine the long-run relationship between poverty, income inequality, and growth on pooled data from eight Houschold Income and Expenditure survey (HIES) datasets compiled between 1992/93 and 2007/08 for the Pakistan. The results show that growth and inequality play significant roles in affecting poverty, and that the effect of the former is substantially larger than that of the latter. Estimation shows that there is a significant positive relationship between inequality and growth in Pakistan. The results at the rural-urban level show that the growth elasticity of inequality is higher in urban areas than in rural areas. Furthermore, growth has a significant positive impact on inequality. The analysis at a regional level shows that both the gross and net growth elasticity of poverty are higher in rural areas than in urban areas, whereas the inequality elasticity of poverty is higher in urban areas than in rural areas. Moreover the results show that growth contributes far more towards reducing poverty, keeping inequality constant, than the latter does to increasing poverty, holding the former constant.

Farhani and Rejeb (2012) investigate the relationship between Energy consumption, GDP and CO2 emissions for 15 MENA countries covering the annual period 1973-2008. The finding of this study reveals that there is no causal link between GDP and EC; and between CO2 emissions and EC in the short run. However, in the long run, there is a unidirectional causality running from GDP and CO2 emissions to EC.

Gani (2012) investigate the relationship between five scope of good governance (political stability, government effectiveness, regulatory quality, rule of law, and corruption) and carbon dioxide (CO2) emissions in a cross-section of developing countries for the time frame 1996 to 2009. By using regression analysis the empirical results present evidence that political stability, the rule of law, and control of corruption are negatively and statistically significantly correlated with CO2 emissions per capita. Moreover results also provide evidence in support of the Environmental Kuznets Curve (EKC) and that trade openness and the size of industrial sector as other strong correlates of CO2 emissions.

Romuald (2012) examines the existence of convergence and the importance of education on carbon dioxide growth per capita, over the period 1970-2004 for 85 countries. By using panel data and apply GMM-System estimation , the results infer the divergence in per capita carbon dioxide emissions around the world, and that education is not a factor in carbon dioxide emissions growth. Furthermore result reveals that, in developed countries, there is convergence for per capita carbon dioxide emissions. Education is found to be a factor in pollution growth, although its effect is mitigated by the presence of political institutions.

Shahiduzzaman and Alam (2012) studied the existence of an inverted U-shape relationship between CO2 emissions per capita and per-capita GDP for the Australia using data from 1961-2009. The empirical result shows the existence of an inverted U-shape relationship between per-capita CO2 emissions and per-capita GDP in Australia.

Boutabba (2013) examines the long-run equilibrium and the direction of a causal relationship between carbon emissions, financial development, economic growth, energy consumption and trade openness for India in a multivariate framework for the time period 1970 to 2008. The empirical results show that there is strong evidence on the long run and causal relationships between per capita carbon emissions, per capita real income, the square of per capita real income, per capita energy use, financial development and trade openness. The results also confirm the existence of EKC hypothesis in the Indian economy. Further, causality tests also indicate that there was a unidirectional Granger causality running from per capita real income, per capita energy consumption, and financial development to per capita carbon emissions.

The above discussion confirms the strong correlation between growth, inequality and poverty in Pakistan. In the subsequent section, an action

has been made to find the statistical relationship between growth-inequality-poverty triangle and existence of environmental Kuznet curve in the context of Pakistan.

DATA SOURCE AND METHODOLOGICAL FRAMEWORK

The study uses annual observations for the period of 1980 to 2011. The data is obtained from various issues of Economic Survey of Pakistan and World Development Indicators which is published by World Bank (2012). The national poverty level is estimated based on an official poverty line of 2,350 calories per adult equivalent per day. To get reliable estimates of poverty and income inequality measures, a simple interpolation technique is used to take the growth in trend between two points in time and fill the data gaps between successive observations. Meanwhile forward interpolation technique is used for the year 2006 onward.

The time series data often shows the property of non-stationarity in levels and the resulted estimates usually provide spurious results (Granger, 1981). Thus, the first step in any time series empirical analysis is to test for the presence of unit roots to remove the problem of inaccurate estimates. The other important step taken was to check the order of integration of each variable in a data series in the model to establish whether the data under hand suffer unit root and how many times it needed to be differenced to gain stationarity (Yousaf et al, 2008).

Cointegration Test

The test for co-integration consisted of two steps: first, the individual series were tested for a common order of integration. If the series are integrated and are of the same order, it implies co-integration. Dickey and Fuller (1979) devised a procedure to formally test for non-stationarity. The Augmented Dickey Fuller (ADF) test was used to test the stationarity of the series. The ADF test is a standard unit root test: it analyses the order of integration of the data series (Dickey and Fuller, 1981). These statistics are calculated with a constant, and a constant plus time trend, and these tests have a null hypothesis of non-stationarity against an alternative of stationarity.

In order to apply Co-integration, we used Johanson's Cointegration test to the series of same order to determine the long run relationship between the variables. If series are cointegrated of order 1, trace test (Johansen's Approach) indicates a unique cointegrating vector of order 1 and hence indicates the long run relationship. In the multivariate case, if the I(1) variables are linked by more than one co-integrating vector, the Engle–Granger (1987) procedure is not applicable. The test for co-integration used here is the likelihood ratio put forward by Johansen and Juselius (1990), indicating that the maximum likelihood method is more appropriate in a multivariate system. Therefore, this study has used this method to identify the number of cointegrated vectors in the model. The Johansen and Juselius method was developed in part by the literature available in the field and reduced rank regression, and the co-integrating vector 'r' is defined by Johansen as the maximum Eigen-value and trace test or static, there is 'r' or more co-integrating vectors. Johansen (1988) and Johansen and Juselius (1990) proposed that the multivariate co-integration methodology could be defined as:

Model 1: CO_2 Emission and Economic Variables

(CO2)t = f (GDP, GINI,POVERTY) (1)

Where CO_2 represents carbon dioxide emissions in kilo ton; GDP represents economic growth in current US $ million; GINI represents income inequality which is measured in percentage; POV represents poverty which is measured by head count ratio in percentage.

Model 2: GDP and Economic Variables

(GDP)t = f (CO_2, GINI,POVERTY) (2)

Model 3: GINI and Economic Variables

(GINI)t = f (GDP, CO_2, POVERTY) (3)

Model 4: Poverty and Economic Variables

(POV)t = f (GDP, GINI, CO_2) (4)

Model 5: Environmental Kuznet curve

(CO2)t = f(GDP, GDP ,GINI, GINI, GDPu GINI) (5)

For the vector of elements representation of model consider the following autoregressive representation:

$$CO2_t = \pi_\circ + \sum_{T=1}^{K} \pi_i (CO2)_{t-1} + \mu_t$$

$$GDP_t = \pi_\circ + \sum_{T=1}^{K} \pi_i (GDP)_{t-1} + \mu$$

$$GINI_t = \pi_\circ + \sum_{T=1}^{K} \pi_i (GINI)_{t-1} + \mu$$

$$POV_t = \pi_\circ + \sum_{T=1}^{K} \pi_i (POV)_{t-1} + \mu$$

Johansen's method involves the estimation of the above equation by the maximum likelihood technique, and testing the hypothesis Ho; $(\pi = \Psi\xi)$ of "r" co-integrating relationships, where r is the rank or the matrix is the matrix $\pi(0\angle r\angle P), \Psi$ of weights with which the variable enter co-integrating relationships and ξ is the matrix of co-integrating vectors. The null hypothesis of non-cointegration among variables is rejected when the estimated likelihood test statistic

$\phi_i \{= -n \sum_{i=r+1}^{p} \ln(1 - \hat{\lambda}_i)\}$ Exceeds its critical value. Given estimates of the Eigen-value $(\hat{\lambda}_i)$ the Eigen-vector (Ψi) and the weights we can find out whether or not the variables in the vector (CO2t) are co-integrated in one or more long-run relationships among three simultaneous equations i.e., Economic Growth (GDP), GINI (income inequality) and Poverty in context of Pakistan.

Error Correction Model (ECM)

If time series are I(1), then one could run regressions in their first differences. However, by taking first differences, we lose the long-run relationship that is stored in the data. This implies that one needs to use variables in levels as well. Advantage of the Error Correction Model (ECM) incorporates variables both in their levels and first differences. By doing this, ECM captures the short-run disequilibrium situations as well as the long-run equilibrium adjustments between variables. ECM term having negative sign and value between "0 to 1" indicates convergence of model towards long-run equilibrium and shows how much percentage adjustment takes place every year.

In the error-correction form, we can effectively determine the separate influences of the short-run and the long-run on CO2 emission. The coefficient of CO2 is expected to be positive in the relationship with GDP. The rationale behind the positive relation is generally accepted that when the GDP increases in chain of outcome the process of industrialization increases, capital investment increases, and access of more resources, high consumption, and direct use of energy. So, all these factors lead to high consumption of fuel (petroleum, Gas, oil) which as a result emits high degree of CO2 emission (Attari and Attaria, 2011).

Secondly, the Income inequality is key variable that determines the income disparity and thus have positive relation with CO2 emission. It means the high income population will directly emit CO2 because the rich people have high marginal propensity to emit CO2. And low income people directly use natural resources to deplete environment and due to less income they cannot take protective measure to save natural environment (Selden and Song, 1994). Whereas, there is another consensus about the income inequality that low income population have less access of resources and low consumption and investment power so they emit low CO2 emission. Furthermore, it is of view that in developing nation the expected sign of CO2 and income inequality is positive as developing nation has high income inequality which leads to high CO2 emission and Developed nations are with positive relation with Income inequality indicating that developed nations have low income inequalities and have low CO2 emission,. More specifically in terms of inequalities, High income inequalities lead to high CO2 emission thus positive relation and Low income inequalities have low CO2 emission therefore having positive sign (Stern et al., 1998). There is also evidence that developing nation have high income inequalities than developed nations because of many factors which contributes for income inequalities such that lack of Government unemployment funds, high pensions, sick compensations, part time job opportunities, basic primary educational target, and incremental high average wage rate etc. (Taylor, et al 2012).

The expected sign for Poverty and CO2 is also positive. The rationale behind this relation is that poor people are direct emitters of CO2 via a

channel of using Direct Natural resources intensively, taking low remedial measures, and due to low education they cannot understand the importance and efficient use of Natural , Economic as well as environmental resources. As a general consensus via a channel of GDP it is believed that the increase in GDP initially increases the income inequality due to rapid production and growth, which then high income inequality leads to poverty. Thus the main reason behind increasing factors of high CO2 emission is poverty (Heshmati, 2006).

Moving towards GINI and GDP, there is complex relation between these variables. As there can be positive, negative as well as no strong relation. The understanding behind positive relation is that with increase in GDP generally in developing nations there are high income inequalities at initial stages of Economic development of nation (Cheema and Sial, 2012). It is also analysed that although increase in GDP decreases poverty but the level of income inequality remains constant due to strong political dominance in the nation and policies which are taken and ruled for the favour of political solidarity (Borghesi, 2000). In case of negative relation between GDP and income inequality the justification reveals that when GDP increases this leads to change the income inequality level and thus lowers the poverty gap, if the Government priorities are not in favor of benefiting poor from growth then the change in poverty gap would be offset by the negative effect of income inequalities. Thus, here although magnitude of GDP reduces the poverty but could not decreases the income inequality (Gelaw, 2009). Moreover the literature also identifies credit market imperfection in low-income developing countries as the likely reason for a strong negative relationship between income inequality and economic growth. While in short run the relationship between growth and income inequality might be positive but over time more income inequalities reduces economic growth. Likewise, it is evidence that more physical and human capital investment and higher government spending have statistically significant positive impact on enhancing economic growth and reducing inequality (Tabassum, 2004).

Furthermore there is negative relationship between GDP and poverty. This means that with the increase in GDP the production increases which then leads to create job and business opportunities. Thus by employment, business trend and increase in per capita income, reduces poverty (Zaman et al, 2011). When the GDP increases the population whether they are subsistence farmers, salaried workers, or self-employed entrepreneurs, poor people derive most of their income from work. This basic fact means that the level of employment, the quality of jobs, and the access which the poor have to decent earnings opportunities will be crucial determinants of poverty reduction. The intuition that jobs matter for development has not been lost on the governments of low income countries and the vast majority of national development strategies look to employment generation as a major channel for poverty reduction (Hull, 2009).

Finally, there is positive relation between GINI and poverty as this relation depends on the conditions. The rationale behind the positive relation is justified in context of level of development and Government policies. The positive relation means with increase in GINI –income inequality the poverty increases and increases in poverty gap increases income inequality as well. It indicate for developing nations, as low GDP growth rate leads to high income inequalities and increases poverty gap so have positive relation with income inequality and poverty. For developed nations as they have high GDP rate which result in low income inequalities reduces poverty gap and boosts economic growth thus in economic chain assigns negative relation between income inequality and Poverty (Stevans and Sessions, 2008).

On the other hand in context of Government polices if targeted towards low income people, basic primary and agricultural education, this would promote the basic income level of poor people by employment and thus lowers the income inequality and poverty ascertaining positive impact and relation as well in income inequality and poverty (Taylor, et al 2012).

Globalization on another fact may also widen inequality. A first channel through which this may happen is off shoring. The tasks that are relocated from richer to poorer countries are typically not skill intensive from the perspective of the skill-rich country, but they are from the perspective of the skill-poor country. As a result, off shoring makes labor demand more skill intensive in both poorer and richer countries, thus increasing inequality in both groups of countries (Feenstra and Hanson, 1996). Second, if firms differ in their profitability and low-income workers work disproportionately in low-productivity firms that are battered by import competition, trade may increase labour income inequality by lowering employment or the relative earnings of low-income workers (e.g. Egger and Kreickemeier, 2009; Helpman et al., 2010). This implied positive link between globalization and inequality is supported by a growing body of studies of individual firms, but it is more difficult to establish a robust link at the aggregate level. Globalizations and technological change may also reinforce each other, thereby further raising inequality. On the one hand, technology may underpin globalization and on the other, the increased competition that comes with globalization may force firms to innovate. Innovation may raise labour income inequality both temporarily since R&D is skill intensive (Dinopoulos and Segerstrom, 1999; Neary, 2003) – and permanently, provided it results in skillbiased technological change (Acemoglu, 2002).

RESULTS AND DISCUSSIONS

Time-series data is often found to be non-stationary, containing a unit root. Ordinary Least Squares (OLS) estimates are efficient if variables, included in the model are stationary of the same order. Therefore, first it is needed to check the stationary of different variables, which are used in our study. For this purpose we applied Augmented Dickey-Fuller (ADF) test. Table 1 gives the results of ADF tests. Based on the ADF tests, all variables are non-stationary at level, but after taking first difference, it would become stationary. Thus, it was concluded that these variables are integrated of order one i.e. I (1).

Table1: Augmented Dickey-Fuller (ADF) on the levels and on the First Difference of the Variables (1980-2011).

Variables		**Level**		**First Difference**		
	Constant	**Trend & Intercept**	**None**	**Constant**	**Trend & intercept**	**None**
CO2	4.559	3.180	9.228	-5.824*	-6.3105*	- 6.955*
GDP	-1.820	-0.895	- 1.412	-5.901*	-5-009*	- 5.962*
GINI	-1.788	-2.035	1.041	-2.762***	-3.351***	- 2.744*
POV	-2.548	-2.539	- 0.623	-3.362**	-3.302***	- 3.412*

Note: *, ** and *** indicates significance at1%, 5% and 10% respectively.

The relationship between the growth-inequality-poverty and carbon emissions was examined by the Multivariate Cointegration Methodology which was proposed by Johansen & Juselius (1990). The results show that there is only one cointegration relationship between the variables GDP, GINI and POV. In another test where we used interaction term, the results indicate that there are three cointegration equations between the variables CO_2, GINI, POV, GDP, GDP square, GINI square and interactive term (GDP u GINI). The presence of cointegration vector shows that there exists a long-run relationship among the variables*. Table 2 shows the short-run error correction model.

Table 2: Short-run Error Correction Model.

Depende nt variables	**Dlog(CO_2)**	**Dlog (GDP)**	**Dlog (GINI)**	**Dlog (POV)**
Constant	0.058* (11.2)	0.215* (2.899)	0.008 (1.440)	-0.018 (-0.627)
Dlog(CO2)	--------	-4.420* (-3.199)	-0.053 (-0.946)	0.070 (0.257)
Dlog(GDP)	-0.01* (-9.80)	---------	0.010** (2.296)	-0.057* (-3.000)
Dlog(GINI)	0.090 (0.294)	3.070 (1.242)	-----------	1.625** (2.207)
Dlog(POV)	-0.07 (-0.95)	-2.146* (-4.003)	0.08*** (1.93)	-----------
R Square	0.458	0.559	0.497	0.485
Adjusted R square	0.323	0.491	0.420	0.402
F-Statistics	5.175*	8.250*	6.441*	5.886*
D.W	1.856	2.282	1.821	1.872

Note: * and ** denotes rejection of hypothesis at 1% and 5% significance level. T- values in bracket.

The results of Table 2 show that there is a negative relationship between carbon emissions and economic growth; however, one percent increase in economic growth reduces poverty around 2.146%. The result implies that CO_2 and other dangerous gases in environment makes the productive area as a red zone where production is impossible or have no economic efficiency or even have to bear high cost to make polluted environment in productive mode. This result is consistent with the study of Burnet (2009) and Tiwari (2011). In addition, one percent increase in poverty, income inequality increases by 0.089 percent. The result is consistent with the study of Tabassum (2004) and Leight (2010). The rationale behind this relation is that economic growth boosts the process of industrialization, increases employment and business opportunities; investment leads to increases the income generation which in chain increases the human capital investment and then increases the per capita income which then ultimately decreases the poverty (Stevans and Sessions, 2008). High income inequality seizes all the benefit of high economic growth and income per capita falls behind the average income. This leads to unemployment and lack of employment, business opportunities, lack of Government funds and benefits increase the poverty. The result indicates that increase in income inequality heavily influence the poverty. The empirical results, given in Table 1, appear to be very good in terms of the usual diagnostic statistics. The value of R adjusted indicates that 32.3%-49.1% variation in dependent variable has been explained by variations in independent variables. F value is higher than its critical value suggesting a good overall significance of the estimated model. Therefore, fitness of the model is acceptable empirically. The result suggests that all variables have a correlation proving the hypothesis. Table 3 shows the long-run error correction term.

Table 3: Long-run Error Correction Model.

Dependent variables	**CO2**	**GDP**	**GINI**	**POV**
Constant	0.0562* (11.977)	0.1973* (2.779)	-0.0079** (-2.443)	-0.004 (-0.238)
Dlog(CO2)	-------------	-4.542* (-4.099)	0.244* (4.354)	-0.183 (-0.542)
Dlog(GDP)	0.0331* (3.601)	-----------	-0.002 (-0.640)	-0.087* (-3.746)
Dlog(GINI)	0.2169 (0.711)	4.524** 2.317)	--------------	2.386** (2.327)

Dlog(POV)	-0.151*** (-1.928)	-1.124** (-2.437)	0.032 (1.198)	-------------
ECM (p)	-1.164* (5.212)	-1.069* (5.010)	0.789* (6.687)	-0.094 (-1.245)
R Square	0.371	0.898	0.718	0.564
Adjusted R 2	0.235	0.876	0.657	0.469
F-Statistics	2.724** (2.821)	40.850* (11.852)	11.764* (7.252*	5.962* (4.898)

DW 2.058 2.166 2.011 1.982

Note: *, ** and *** denotes rejection of hypothesis at 1%, 5% and 10%significance level. T- values in bracket.

The results of long-run error correction model indicate that the adjustment parameter (p) shows negative value, indicating the long-run convergence. The ECM estimation reveals that in between 78.9% to 116.4% disequilibrium on dependent variable is reported with other independent variables. The large sizes of coefficient of error-correction terms indicate that speed of adjustment is too fast for equations to return to their equilibrium level, once it has been shocked. The model extends an EKC to allow for an income inequality effect. The study used interaction effect which shows the relationship between CO2 emissions per capita and income inequality to depend on the value of GDP per capita. The results reveal that the relationship between carbon dioxide emissions per capita and income inequality is U-shaped i.e., for Pakistan characterized by high income inequality, reductions in income inequality are associated with lower per capita emissions. For less unequal societies, reductions in income inequality are associated with increases in carbon emissions per capita. The inequality effects are highly significant and thus clearly provide a better fit of the data than a linear effect.

The level at which reductions in income inequality stop being beneficial would depend on the level of GDP per capita. The coefficient estimates for specification is found in Table 4. The higher the economic growth in initial stage increases the emission but with the increase in

GDP gradually at certain the emission is at peak level, then again in third stage of development the CO2 emission decreases. The result shows that the relationship between income inequality and per capita carbon emissions is negative and GDP^2 is with correct sign and is significant. Moreover, the value of GINI is also significant. Similarly there is negative and significant relation between interaction term and CO_2 emission. In particular, the findings are consistent with the aggregation bias argument and a more complex relationship between income and the MPE. For example, if there is a section of low incomes where the MPE is zero, as people are outside of the carbon economy, then rises, then falls, and rises again at very high levels of incomes, this could deliver the results that found here, including the different turning-points for stage of economic development. The Suggestive descriptive evidence supports the correlation between income inequality, CO_2 emission, and GDP .Thus in poorer countries higher inequality reduces access and use of these goods, while in richer countries it increases it, confirming the supposition that the poor in poor countries are largely outside of the carbon economy while in rich countries, higher incomes might be associated with a rising MPE. Finally the goodness of fit model shows that 90.4 percent changes in CO2 emission is due to change in explanatory variables such that GDP, GINI, GDP^2, $GINI^2$ and interaction term (GDPXGINI).

Table 4: Existence of Environmental Kuznet Curve.

Dependent variable : CO2	Coefficient(t-value)
Constant	986295.6* (3.608)
GDP	(33950.29)* (2.830)
GDP^2	-179.639 (-0.017)
GINI	-58544.88* (-4.364)
$GINI^2$	(866.143)* (5.217)
GDPxGINI	-825.573* (-2.955)
POV	(-430.152) (-1.677)

Statistical Tests:

R-squared: 0.904; Adjusted R- squared: 0.881; F-statistics: 39.547*; Durbin Watson: 1.864

Note: * denotes rejection of hypothesis at 1% significance level. T- values in bracket.

CONCLUSION AND POLICY RECOMMENDATIONS

Environmental degradation is not a new thing, it has been happening all over the world for centuries. The problem is that it is now occurring at a much faster rate, therefore, not leaving enough time for the environment to recover and regenerate. The greater demands placed on the environment by an ever increasing human population is putting a great strain and drain on the earth's limited natural resources (SEEN, 2012). Environment is very indispensable in every aspect of life. All the living beings are influenced by the environment. All the components required for survival of living organisms are gained through environment containing air, water, soil, food, clean environment etc. The development of living creatures i.e., quality of human life, sector developmental activities, growth, natural resources for development of civilization of human beings, even plant and animals are also depend upon the environment. Therefore, all the process of entities is done under the environment (Pilisuk, et al 1996).

The current issues of environment are open up to many categories such that water pollution from raw sewage, industrial wastes, and agricultural runoff; limited natural freshwater resources; most of the population does not have access to potable water; deforestation; soil erosion; desertification, Green houses gases and CO2emission (CIA, 2012). Environmental pollution has been rapidly increasing day by day. The vehicles, industries, power stations produce smoke, dust and harmful effects on human beings. Pollution can cause diseases like tuberculosis, cancer, disease of lungs and skin to the people. Water is polluted by sewage, chemicals from industries and factories, garbage and other biodegradable materials that can make the water undrinkable. There are different types of water born disease like cholera, typhoid, jaundice, etc. The noise that comes from vehicles and crowed has negative effects on human health causing hypertension problems (Krieger and Higgins, 2002). The soil pollution due to the excessive use of fertilizers causes barrenness of soil which is responsible for lowering plantation- forestation and increases CO2 emission in large extent. Thus all these kinds of pollution are the factor for unhealthy environment (Dipshikha, 2010).

Carbon emissions, most notably carbon dioxide (CO2), are part of a collection of gases that negatively influence the quality of our air and increase the greenhouse effect. Greenhouse gases have a direct influence on the environment, causing extreme weather changes, a global temperature increase, the loss of ecosystems and potentially hazardous health effects for people (McMichael et al, 2003). Low Carbon Economy (LCE) is the today reality, which we can't hinder. The CO2 is adversely affecting the environment which is against the sustainable goals. Now, the relationship between the energy consumed and economic growth is mainly concerned while investing in the power generation sector. The race of being the Asian power is not out of sight in the region. Low carbon technologies are the best key for the development of the world economy. Through diversification strategy for the energy, which consist of three parts: the energy resources, energy production and end use of energy can be useful in sustaining environment (Ighodaro – CAU, 2010). The essence of the diversification of energy resources is to reduce the coal consumption ratio in the total energy consumption; to fast develop the natural gas industry; to make full use of domestic energy, and reasonable use of international ones; and to exploit the hydroelectric, nuclear, and renewable energy, for the realization of diversified energy supply pattern (Xie, 2009).

Moving towards the income inequality empirically, econometric analysis in this research shows that the income inequality affects negatively environmental quality and environment degradation worsens population's health and significantly emits high CO2 emission. So, this negative effect of income inequality on environment quality is mitigated by good institutions (Drabo, 2010). Confirming that the environment quality is an important channel through which income inequality affects population health. As policy implication, it is suggested that income inequality is bad for health and environment, and countries with high income inequality may implement distributive policy in order to avoid its negative impact on health (Hunter, and Amanda, 2004).

Last but not least the economic variable - poverty contributes large proportion of CO2 emission. It is often the people living in degraded environments are responsible for the damage that as occurred, but this is often as a result of underlying social and economic problems (SEEN, 2012). Poverty is a major contributing factor to land degradation as it forces millions of people to destroy the resources that are around them in order just to survive. Poor people often do not have access to the best land, leaving them to depend on the most fragile areas and resources. Their situation may mean that they have no other choice other than to use what resources are available to them, even if these result in the degradation of the land (Shah, 2001). Environmental Degradation can be copped by Bio-physical procedure, (Planting trees over degraded land or afforestation can help to protect the soil from strong wind and from being washed away by soil erosion), Soil salinization and waterlogging can both be avoided by using drip irrigation which delivers only as much water as is needed to a specific area as opposed to high-pressure sprinklers which cause the soil to be permanently covered with water (Verbeten, 1998). The mass effects of poverty can only be eliminated through the agreement and implementation of the World's leading politicians. Without their commitment to end debt, make trade rules fairer, and give more aid to those countries in need poverty will not be eliminated. Politicians also need to sincerely address the pressing environmental problems that face our societies today, and again commit to policies that enforce the sustainable use of our planet (Speth, 2008).

The rising inequality and poverty is significantly increasing day by day. Policy makers must therefore devote more attention to inequality. The following measures can be taken to alleviate poverty and income inequalities:

- redistributive transfers;
- investment in universal access to health and education; x progressive taxation;
- Removal of the barriers to equal rights and opportunities for women.
- Reforming land ownership, ensuring the right access to land and other resources, and investing in small-scale food producers (SEEN, 2012).

Federal government mandates should take remedial measures regarding control of carbon emissions. It can be done directly by a carbon tax, which is exactly sound strategy -- taxing companies, factories, industries directly, based on the amount of carbon they put into the atmosphere, and efficient fuel consumption by transport sector. The goal of a carbon tax would be to convince businesses and other organizations to reduce their total emissions (Pelosi, 2009) Secondly, the governmental control approach that has been under study in recent years is referred to as cap-and-trade legislation. In this system, the government sets a “cap” on the maximum amount of emissions it will allow. From here, it then auctions off emissions allowances to companies until it reaches that cap. Companies that cannot cover their emissions with their allowances are forced to either reduce their totals or buy allowances from other companies. This system is designed to promote stricter emissions standards without directly taxing companies (Wall Street Journal, 2010).

Currently United Nations Development Programme (UNDP) is working towards providing support in the implementation of the national environmental agenda. The Environment and Climate Change Unit works on improving environmental management through capacity building of environmental institutions; mainstreaming environment into the development processes; implementation of innovative solutions for meeting the challenges posed by climate change and sustainable use of natural resources; and advocacy (UNDP, 2010). UNDP engages national, provincial and local partners to obtain environmental information for decision-making and enhancing global and regional environmental cooperation. This also serves to advance implementation of national and regional environmental activities and create linkages among stakeholders and government for policy and implementation. UNDP also scales up coordination efforts to anchor international assistance at the provincial level (UNDP, 2010).

Now the aim of Government of Pakistan should promote sustainable use of natural resources and helps institutions understand and effectively use environmental management strategies for ecosystems, land, water, soil and bio-diversity (FAO, 2011). The appropriate measures and adapting management systems along with the empowerment at grassroots organizations is crucial need of environmental management. To help mitigate food, water and energy insecurity, the policy maker should help to build resilience to climate change. This is done by contributing to climate change adaptation and mitigation interventions, through partnership and technology transfer with a focus on youth, women and vulnerable groups (UNDP, 2010). The Government and ruling agencies should strengthen the Initial Environmental Examination and Environmental Impact Assessment components of the crisis response programmes to undertake rapid assessments of environmental damage (Economic Survey of Pakistan, 2010).

REFERENCES

1. Acemoglu, D. (2002), Technical Change, Inequality, and the Labor Market. Journal of Economic Literature, Vol. 40, No. 1, pp. 7-72. Online available at: http://www.aeaweb.org/articles.php?doi=10.1257/0022051026976
2. Acharray, J (2009). FDI, Growth and The Environment: Evidence from India on CO2 Emission during the last two decades. Journal of Economic Development. Vol.43, Issue No: pp.1-16.
3. Ahmed, M (2001). Estimation of Distribution of Income among various occupations/Professions in Pakistan. Pakistan Economic and Social Review Volume. XXXIX, No. 2 , pp. 119-134
4. Ajl, M (2010). New Study Measures Total Environmental Degradation, Not Just CO2 Pollution. Inside Climate News. Online available at:
5. http://insideclimatenews.org/print/4606
6. Alkhathlan, K, Alam, M, Q and Javid, M. (2012). Carbon Dioxide Emissions, Energy Consumption and Economic Growth in Saudi Arabia: A Multivariate
7. Cointegration Analysis. British Journal of Economics, Management & Trade. Vol. 2, Issue No.4, pp. 327-339, online available at: www.sciencedomain.org.
8. Andrich, M. A., Imberger, J., & Oxburgh, E. R. (2010). Raising Utility and Lowering Risk through Adaptive Sustainability: Society and Wealth Inequity in Western Australia. Journal of Sustainable Development, 3(3), 14-35.
9. Ang, J. B. (2007): CO2 emissions, energy consumption, and output in France. Energy Policy, 5, 4772-4778.
10. Anietie Vincent essien (2011). The relationship between economic growth and co2 emissions and the effects of energy consumption on C02 emission patterns in Nigerian economy. pp.1-35. Online available at: http://ssrn.com/abstract=2148916.
11. Annual Energy Review (2004). U.S. Department of Energy, Energy

Information Administration. Online available at : http://www.whatsyourimpact.eu.org/co2sources.php

12. Apergis N. and Payne J.E. (2009): CO2 emissions, energy usage and output in Central America. Energy Policy 37 (2009) 3282–3286
13. Arouri,M,E,H ., Ben,A., M'henni,H., Rault,C., (2012). Energy Consumption, Economic Growth and CO2 Emissions in Middle East and North African Countries." IZA Discussion Paper No. 6412. Online available at: http:/www.ideas.repec.org/a/eee/enepol/v45y2012icp342-349.html.
14. Attari, M, I ,J and Attaria,S,M. (2011). The Decomposition Analysis of CO2 Emission and Economic Growth in Pakistan India and China. Pakistan Journal of Commerce and Social Science. Vol. 5, Issue No: 2, pp.330-343.
15. Boopen, S and Vinesh,S (2010). On the Relationship between Co2 Emissions and Economic Growth: The Mauritian Experience. Online available at:
16. www.csae.ox.ac.uk/conferences/2011-EDiA/.../776-Seetanah.pdf
17. Borghesi, S.(2000). Income Inequality and the Environmental Kuznets Curve. Online available at: www.feem.it/userfiles/attach/Publication/NDL2000/NDL2000083.pd
18. Boutabba, M,A. (2013). The impact of financial development, income, energy and trade on carbon emissions: Evidence from the Indian economy. pp.1-21. Online available at: http://www.epee.univ-evry.fr/RePEc/eve/wpaper/epee-2013.rdf.
19. Boyce, J. K., Klemer, A. R., Templet, P. H., & Willis, C. E. (2007). Power distribution, the environment, and public health: A state-level analysis. Ecological Economics. 29(2007), 127–140.
20. Burnett,J,W(2009). Economic Growth and Environmental Degradation. Online available at:
21. http://www.ageconsearch.umn.edu/bitstream/46838/2/Burnett_manuscript1.pdf
22. Cheema,A,R., and Maqbool H. Sial,M,H.,(2012). Poverty, Income Inequality, and Growth in Pakistan: A Pooled Regression Analysis. The Lahore Journal of Economics. Vol. 17, Issue No. 2, pp. 137–157
23. CIA (2012). The world FactBook- Environment - current issues.

Online available at: https://www.cia.gov/library/publications/the-worldfactbook//fields/2032.html.

24. Dinopoulos, E. and P. Segerstrom (1999). A Schumpeterian Model of Protection and Relative Wages. American Economic Review, Vol. 89, No. 3, pp. 450-472.
25. Dipshikha (2010) News article consequences of pollution. Online available at: http://enviroment-dipshikha.blogspot.com/2010/08/consequences-of-environmentalpollution.html
26. Drabo, A. (2010). Impact of Income Inequality on Health: Does Environment Quality Matter? Online available at: www.csae.ox.ac.uk/conferences/2010EDiA/papers/114-Drabo.pdf
27. Dulal, H. B., Foa, R., & Knowles, S. (2011). Social Capital and Cross-Country Environmental Performance. Journal of Environment & Development, 20(2), 121– 144.
28. Economic Survey of Pakistan (2010). Global HDR Report- 2010. Online available at http://hdrstats.undp.org/countries/country_fact_sheets/cty_fs_PAK.html) Egger, H. and U. Kreickemeier (2009). Firm Heterogeneity and the Labour Market Effects of Trade Liberalization. International Economic Review, Vol. 50, No. 1, pp. 187-216.
29. FAO (2011). Biodiversity for Food and Agriculture Contributing to food security and sustainability in a changing world. The Food and Agriculture Organization of the United Nations and the Platform for Agrobiodiversity Research.pp.1-78, online available at: www.fao.org/fileadmin/templates/**biodiversity**.../PAR-FAObook_lr.pdf .
30. Farhani, S and Rejeb, J, B (2012). Energy Consumption, Economic Growth and CO2 Emissions: Evidence from Panel Data for MENA Region. International
31. Journal of Energy Economics and Policy. Vol. 2, Issue No. 2 pp. 71-81.ISSN: 2146-4553. Online available at: www.econjournals.com
32. Feenstra, R.C. and G.H. Hanson (1996). Foreign Investment, Outsourcing and Relative Wages. NBER Working Paper, No. 5121, National Bureau of Economic Research.
33. Fritz Kahrl,F., and Holst,D,R (2007). Carbon Inequality. Center for energy, resources, and economic sustainability. Research Paper No. 07090701.pp.1-11.

34. Gani, a. (2012). The relationship between good governance and carbon dioxide emissions: evidence from developing economies. Journal of economic development. Volume 37, Number 1, pp.77-93.
35. Gelaw, F (2009). The relationship between poverty, inequality, and growth in the rural Ethiopia: micro evidence. International Association of Agricultural Economists Conference, Beijing, China, pp.1-26.
36. GoP (2010). Economic Survey 2009-10, Economic Advisor's Wing, Ministry of Finance, Government of Pakistan, Islamabad (2010).
37. Grimm, M. (2003). The medium and long term effects of an expansion of education on poverty in Côte d'Ivoire A dynamic micro-simulation study in International Conference on Staying Poor: Chronic Poverty and Development Policy; Institute for Development Policy and Management, 7-9 April 2003, University of Manchester (U.K.), DIAL (Paris).
38. Grunewald, N., Klasen, S., Martínez-Zarzoso,I., Muris,C., (2011). Poverty, Equity and Growth in Developing and Transition Countries: Statistical Methods and Empirical Analysis. Courant Research Centre. Discussion paper No.92.pp.1-8.
39. Heerink,N,, Mulatu,A, Bulte,E. (2001). Income inequality and the environment: aggregation bias in environmental Kuznets curves. Ecological Economics, Vol. 38, pp.359–367.
40. Helpman, E., O. Itskhoki and S.J. Redding (2010). Inequality and Unemployment in a Global Economy. Econometrica, Vol. 78, No. 4, pp. 1239-1283.
41. Heshmati A. (2006). Conditional and Unconditional Inequality and Growth Relationships. Applied Economics Letters 13, 925-931.
42. Hunter,D, J and Amanda, K (2004). Tackling health inequalities: turning policy into practice? Health Development Agency. ISBN 1-84279-229-6. pp.1-28. Online available at: www.who.int/rpc/meetings/Hunter_Killoran_Report.pdf
43. Ighodaro CAU (2010). Co-Integration And Causality Relationship Between Energy Consumption And Economic Growth: Further Empirical Evidence For Nigeria. Journal of Business Economics and Management 11(1):97-111.
44. Jabeen. M (2012). Alleviating Poverty and Preventing Environmental

Degradation. The Information and Public Affairs Department of Muslim Aid, UK. Online available at :http://reliefweb.int/report/world/alleviating-poverty-and-preventing-environmental-degradation.

45. Keskin,A (2002). Income distribution and economic growth: a cross country study. Online available at: http://www.web.inonu.edu.tr/~ozal.congress/pdf/44.pdf. Khan, A, N., Ghauri,B, M., Jilani, R., and Rahman,S., (2009). Climate change: emissions and sinks of greenhouse gases in Pakistan. Pakistan Space and Upper Atmosphere Research Commission (SUPARCO). pp. 146-159.
46. Khanam, F (2007). Primary Education Expansion, Income Inequality and Poverty Reduction. Pp.1-40. Online available at:
47. http://www.ww3.comsats.edu.pk/thesis/ThesisDetail.aspx?id=431
48. Knutsen, C,H (2008). "GDP, Inequality, Democracy and the Happiness of Nations". Online available at: http://www.folk.uio.no/carlhk/publications/SWB.pdf .
49. Krieger, J and Higgins, D, L (2002). Housing and Health: Time Again for Public Health Action. American Journal of Public Health. 92(5): pp.758–768. Online available at: http://www.ncbi.nlm.nih.gov/pmc/articles/PMC1447157/
50. Leight, E,H., (2010). The Relationships Between Inequality, Economic Growth, and Political Control: Explorations Using U.S. State-level Panel Data, 1969 – 2005. Pp.1-120.Onlineavailabl at: http://www.wesscholar.wesleyan.edu/cgi/viewcontent.cgi?article=1443...etd.
51. Leitao, N, C and Shahbaz, M (2013). Carbon dioxide emissions, urbanization and globalization: a dynamic panel data. The Economic Research of Guardian. Vol.3, issue No.1. pp. 22-33. ISSN: 2247-8541.
52. Leitao, N,C (2013). The environmental kuznets curve and globalization: the empirical evidence for Portugal, Spain, Greece and Ireland. Energy Economics Letters, Vol.1, Issue No.1, pp. 15-23. Online available at: http://aessweb.com/journal-detail.php?id=5049
53. M'henni H. (2005): Economic development, adjustment and environmental quality: the case of Tunisia for a Contingent Valuation Study. Mediterranean Journal of Economics , Agriculture

and Environment Management . Vol.4, Issue No, 2. Pp. 147-162.

54. McCallum, R (2012). Poverty and Environmental Degradation. Future Foreign Policy- The university Think tank of UK International Affairs. Online available at:http://reliefweb.int/report/world/alleviating-poverty-and-preventing-environmental-degradation
55. McMichael, A, J., Lendru, D,H,C, Corvalán, C, F, Ebi, K, L , Githeko, A, K., Scheraga, J, D, and Woodward, A (2003). Climate change and human health Risks and Responses. Climate Change And Human Health- Pp.1-33. Online available at: www.who.int/entity/globalchange/publications/climchange.pdf
56. McPherson, M. A., & Nieswiadomy, M. L. (2005). Environmental Kuznets curve: threatened species and spatial effects. Ecological Economics. Issue No: 55, pp.395-407.Online available at: http://dx.doi.org/10.1016/j.ecolecon.2004.12.004
57. Mekenbayeva, K., and Karakuş,S,B., (2011). Income inequality and economic growth: enhancing or retarding impact? Online available at :
58. www.tcmb.gov.tr/yeni/.../kamila_mekenbayeva-semih_baris_karakus.pd.pp.1-37.
59. Monbiot, G (2008). Article Heat- How to Stop the Planet Burning. Online available at: http://www.pbs.org/wgbh/pages/frontline/heat/etc/worldco2.html
60. Murad, W and Mustpha, N (2010). Does poverty cause environmental degradation? Evidence from waste management practices of the squatter and low-cost flat households in Kuala Lumpur. World Journal of Science, Technology and Sustainable Development, Vol. 7, Issue No. 3, pp.275 – 289. ISSN: 2042-5945.
61. Neary, J.P. (2003). Presidential Address: Globalization and Market Structure. Journal of the European Economic Association, Vol. 1, No. 2-3, pp. 245-271.
62. Ozturk I., Acaravci, A.(2011). CO2 emissions, energy consumption and economic growth in Turkey. Renewable and Sustainable Energy Reviews 14(9): 32203225.
63. Papanyotou, T. (1994). Empirical tests and policy analysis of environmental degradation at different stages of economic development. Pacific and Asian Journal of Energy, 4(1), 23-42.

64. Payne, J. (2010): Survey of the international evidence on the causal relationship between energy consumption and growth. Journal of Economic Studies, 37(1), January, 53-95.
65. PBS (2010). Online available at : www.pbs.gov.pk/content/social-statistics.
66. Pelosi, N (2009). The Cap and Tax Fiction-Democrats off-loading economics to pass climate change bill. The wall street Journal. Online available at:
67. http://online.wsj.com/article/SB124588837560750781.html
68. Pilisuk, M, McAllister,J and and Rothman, J (1996). Coming Together for Action: The Challenge of Contemporary Grassroots Community Organizing. Journal of Social Issues, Vol.52.No.1. pp.15-37. Online available at: http://www.sonoma.edu/users/w/warmotha/psychclasses/490/spring01/grassroots.html Poverty, Income Inequality, and Growth in Pakistan: A Pooled Regression Analysis.
69. Ravallion, M (2000). Why Don‘t We See Poverty Convergence? Development Research Group, World Bank 1818 H Street NW, Washington DC, 20433, USA. Online available at: www.cgdev.org/doc/events/3.09.10/Martin_Ravallion.pdf
70. Roca, J. and Alcantara, V (2001). Energy intensity, CO2 emissions and the environmental Kuznets curve. The Spanish case. Energy Policy , Vol . 29, Issue No. 7, pp. 553-556 .Online available at : http://resolver.scholarsportal.info/resolve/03014215/v29i0007/553_eiceatekctsc
71. Romuald,K,S., (2012). Education, Convergence and Carbon Dioxide Growth per Capita. African Journal of Science, Technology, Innovation and Development Vol. 3, Issue No. 1, pp. 65-85.
72. Ru,X. Chen,S., Liu,Y., Su,L.(2012). A Study on Evolution and Driving Forces of Carbon Dioxide Emissions. Journal of Sustainable Development. Vol. 5, Issue No. 5; pp.110.120. online available at : http://dx.doi.org/10.5539/jsd.v5n5p111
73. SEEN (Supporting Environment and Education in NAMBIA) (2012). Environmental Degradation -Seen Environmental Learning Information Sheet no 5. Natural Resources and their Management.pp.1-5. Online available at: www.nied.edu.na/.../

Environmental%20Information%20Sheets/**Natural**%. Selden, T.M., Song, D., 1994. Environmental quality and development: is there a Kuznets

74. Shah, A(2001). Article, Stress on the environment, society and resources? Online available at: http://www.globalissues.org/article/214/stress-on-the-environmentsociety-and-resources
75. Shahbaz, M., Hooi L,H., Shabbir,M,S., (2009). Environmental Kuznets Curve and the Role of Energy Consumption in Pakistan. pp.1-29. Online available at: www.pide.org.pk/psde/25/pdf/agm26/.../Muhammad%20Shahbaz.pdf
76. Shahiduzzaman, Md and Alam, K (2012). Relationship between CO2 emissions and income in Australia: is there any evidence of an EKC? In: 41st Australian
77. Conference of Economists (ACE 2012), 8-12 July 2012, Melbourne, Australia. Online available at: http://ace2012.org.au/ACE2012/11__Conference_Papers/Conference_Proceedings.aspx
78. Sheikh, A,T (2008). Challenge of Climate Change: Pakistan's carbon emissions continue to grow at an increasing rate.(Daily DAWN - February 16, 2008) Smith, S.E (2013). Environmental Degradation. Article from wiseGEEK. Online available at: www.wisegeek.com/what-is-environmental-degradation.htm Speth, J, G (2008). Environmental Failure: A Case for a New Green Politics. Online available at:
79. http://e360.yale.edu/feature/environmental_failure_a_case_for_a_new_green_politics/2075/ Stern, D.I., 1998. Progress on the environmental Kuznets curve?
80. Stern,D,I, Auld,T., Common,M.,S.,. Sanyal.,K.,K (1998) I. There an Environmental Kuznets Curve for Sulfur? Online available at:http//:www.
81. econpapers.repec.org/RePEc:anu:wpieep:9804
82. Stevans,L,K., and Sessions,D,N., (2008). The Relationship between Poverty and Economic Growth Revisited. Journal of Income Distribution, Vol. 17, Issue No.1, pp. 5-20.
83. Tabassum. A (2004). Economic Growth and Income Inequality Relationship: Role of Credit Market Imperfection. Online available at:

84. http://www.pide.org.pk/psde24/pdf/19.pd. pp.1-22
85. Taylor,M., Haux,T., and Pudney,S (2012). Can improving UK skills levels reduce poverty and income inequality by 2020? Inspiring social change. pp.1-60. Online available at: www.jrf.org.uk/sites/files/jrf/skills-poverty-income-inequality-full.pdf
86. Tiwari, A, K (2011). Energy Consumption, CO2 Emission and Economic Growth: Evidence from India. Journal of International Business and Economy. Vol.12, Issue No.1, pp. 85-122.
87. Torras, M. and Boyce, J.K. (1998). Income, inequality and pollution: a reassessment of the environmental Kuznets curve. Ecological Economics. 25, 147-60. Tucker (1995). Carbon dioxide emissions and global GDP. Ecological Economics. Vol.15. pp.215-223. Online available at: www.sciencedirect.com/science/article/pii/0921800995000453
88. U.S. Department of Energy, (2007). Energy Information Administration, Carbon Dioxide Information Analysis Center.
89. U.S. Department of Energy, Energy Information Administration. Manufacturing Consumption of Energy 2002.
90. U.S. Department of State (2007). Fourth Climate Action Report to the UN Framework Convention on Climate Change: Projected Greenhouse Gas Emissions. U.S. Department of State, Washington, DC, USA.
91. UNDP (2010). Empowered lives. Resilient nations. Online available at: http://undp.org.pk/component/option,com_publications/sec_id,7/view, article/
92. Usenobong F. Akpan and Agbai Chuku 2011. Economic Growth and Environmental Degradation in Nigeria: Beyond the Environmental Kuznets Curve. Munich Personal RePEc Archive (MPRA) Paper No. 31241, pp.1-29. Online at http://mpra.ub.uni-muenchen.de/31241/.
93. V. Suri,V and Chapman, D (1998). Economic growth, trade and energy: implications for the environmental Kuznets curve. Ecological Economics. Vol.25. pp.195208.
94. Verbeten, E (1998). Irrigation in arid and semi-arid environments. Interaction between Migration, Land & Water Management and Resource Exploitation in the Oases of the Maghreb. IMAROM Working Paper Series no. 1, pp. 1-64. Online available at:

95. www.heindehaas.com/.../IMAROM%20Working%20Paper%20 01%20.
96. Wall street journal (2010). Green planct- how carbon emission effect environment. Online available at: http://curiosity.discovery.com/question/how-carbonemissions-affect-environment.
97. Wallace, O (2013). What is Environmental Degradation? Wise GEEK. Online available at: http://www.wisegeek.com/what-is-environmental-degradation.htm Wilkinson, R., and Pickett, K. (2010). The Spirit Level: Why Equality is Better for Everyone. New York: Penguin Books Ltd.
98. WWF International. (2010). Living planet report 2010. Gland: WWF International. Retrieved September 29, 2011, online available at:
99. http://wwf.panda.org/about_our_earth/all_publications/living_planet_report/
100. Xie, K. C. (2009). Low carbon economy and energy technologies for a low carbon future. Energy Sources, 31, 1593-1597.
101. Zaman, K., Shah, I,A., Khan.,M,M., Ahmad,M. (2011). Exploring the Link between Poverty-Pollution-Population (3Ps) in Pakistan: Time Series Evidence.
102. Journal of Economics and Sustainable Development. Vol.2, Issue No.11&12. pp. 1-27. ISSN 2222-1700.

CHAPTER 9

THE IMPORTANCE OF CORPORATE BRAND IDENTITY IN BUSINESS MANAGEMENT: AN APPLICATION TO THE UK BANKING SECTOR

Isabel Buil [a], Sara Catalánb, Eva Martínez [b]

[a] Departamento de Dirección de Marketing e Investigación de Mercados, Facultad de Economía y Empresa, Universidad de Zaragoza, Edificio Lorenzo Normante, C/ María de Luna s/n, 50018 Zaragoza, Spain

[b] Departamento de Dirección de Marketing e Investigación de Mercados, Facultad de Economía y Empresa, Universidad de

Zaragoza, Gran Vía 2, 50005 Zaragoza, Spain

ABSTRACT

Corporate brand identity management is a key issue for any organisation. Accordingly, its study is a research field of great interest. This paper seeks to broaden the understanding of this strategic activity and its effects. Specifically, it investigates the concept of corporate brand identity from the employees' perspective in the UK financial banking sector and analyses the link between brand identity management and employees' attitudes and behaviours. Results indicate that organisations should pay special attention to the corporate brand identity management, given its influence on employees' commitment with their organisations, as well as their brand performance and satisfaction.

JEL classification

M3

Keywords: Corporate brand identity, Commitment, Performance, Satisfaction, Banking

INTRODUCTION

Identity is the cornerstone of one of the most important areas of study in recent years: corporate marketing (Balmer, 2008). This term provides the foundation for numerous concepts, such as corporate brand identity (Balmer, 2012). This identity type encompasses specific values and features that, associated to the corporate brand, represent an organisation and the products offered to the market (Balmer and Greyser, 2002).

Corporate brand identity is an effective strategic tool and an important source of sustainable competitive advantages, which provides multiple benefits to the organisations (Melewar, 2003). Its management, understood as a comprehensive view of the different dimensions that need to be internally managed and controlled by the organisation (Simões et al., 2005), is essential as corporate brand identity precedes and represents the basis for brand image. Overall, due to its growing importance, some authors have incorporated this concept as a key element of the new marketing paradigm (Alsem and Kostelijk, 2008).

Despite the relevance of corporate identity management in recent years, there is little research that empirically explores this strategic resource (Arendt and Brettel, 2010; Simões et al., 2005). To date, previous studies have mainly focused on the definition and conceptualisation of this construct. Therefore, several authors advocate further exploration of the empirical impact of its management (Blombäck and Ramírez-Pasillas, 2012).

In addition, research on corporate brand identity management has usually focused on the opinions and perceptions of brand managers, consultants and others experts. Despite the literature has largely acknowledged the critical role of employees in brand success, few works have adopted the perspective of employees in their studies (Burmann et al., 2009; Malhotra and Mukherjee, 2004). This issue is specially important in the services sector, as employees embody the services brands and their performance brings brand promises to life (Punjaisri and Wilson, 2011; Wallace et al., 2013).

Within the services sector, the study of corporate brand identity management is specially relevant in the banking sector, which has been notably damaged by the economic and financial crisis. In addition to the financial problems, the banking sector has recently faced diminishing consumer trust and confidence (Colton and Oliveira, 2009). In this sense, a survey developed in the UK showed that a high percentage of customers, around 40%, were totally disengaged with their banks, whereas a 69% asserted that there are no clear differences among the banks (McEwen, 2010). In this context, it is critical to restore customers' trust and confidence, trying to convey a positive and coherent image with the essence of the organisation. It is in this complex process where corporate brand identity management and the role played by employees are crucial.

The aim of this research is, therefore, to gain a better understanding of the effects of corporate brand identity management. To reach this objective, corporate brand identity management is explored from the employees' perspective in a sector of great interest, the UK banking sector. In addition, the research analyses the relationship between corporate brand identity management and employees' attitudes and behaviours. Specifically, it assesses the effect of identity management on employees' commitment to the organisation, as well as the impact that such commitment has on their brand performance and satisfaction.

This research contributes to the literature in several ways. First, it extends previous research by providing empirical evidences on the study of corporate brand identity management and adopting the perspective of employees. Furthermore, the findings of this study offer useful guidelines on how to manage the corporate brand identity and provide helpful implications for banks and other organisations.

The paper is organised as follows. It opens with a brief, general discussion of the corporate brand identity concept. The next section dcvclops thc conceptual framework and the research hypotheses. This is followed by the methodology and the analysis of empirical findings. The paper then outlines the conclusions and implications of the research. Finally, it addresses the limitations of the study and identifies directions for further research.

Corporate brand identity

The term identity is the basis for a great number of interrelated concepts, such as brand identity, corporate identity or organisational identity, among others (Currás, 2010). The present research focuses on the concept of brand identity, and more specifically, of the corporate brand, which is very common in the services sector.

Brand identity has been studied from a wide range of disciplines. This multidisciplinary nature has led to a variety of definitions and conceptual frameworks that reflect the complex and multidimensional character of this relevant construct. Overall, brand identity includes a set of features and dimensions that determine the brand's way of being, thinking and behaving. It is, therefore, a key concept in brand management, as brand identity defines not only the purpose and meaning of the brand, but also the directions to follow.

Table 1 shows some of the most important brand identity definitions. Several aspects can be highlighted from these definitions. On the one hand, scholars in brand management define brand identity as a unique set of aspects of the brand, as well as the idea or essential meaning associated to it. Similarly, all of them emphasise the internal nature of this construct, which emanates from the organisation. This internal vision is, however, questioned by numerous scholars. In this sense, although emanating from brand managers, brand identity is further developed by other actors, such as employees and consumers (da Silveira et al., 2013). Branding literature has also discussed the stability of brand identity over time. In this sense, although brand identity definitions mainly emphasise that brands should maintain their identity, others reconceptualise brand identity as a dynamic concept. As such, it is possible to distinguish a core identity, comprising the central and timeless essence of the brand and an extended identity, which includes other dynamic dimensions that may change in light of different contexts (Aaker, 1996; da Silveira et al., 2013).

Table 1: Brand identity definitions and dimensions of the main conceptual frameworks.

Author	Definition	Dimensions
Aaker (1996)	A unique set of brand associations that the brand strategist aspires to create or maintain	Twelve dimensions organised around four perspectives: - Brand as a product (product scope, product attributes, quality/value, uses, user, country of origin) - Brand as a person (brand personality, brand–customer relationships) - Brand as an organisation (organisational attributes, local vs. global) - Brand as a symbol (visual imagery/metaphors, brand heritage)
Kapferer (2004)	A brand's meaning as put forward by the firm	Brand identity prism: - Physique: set of the brand's physical features evoked in people's minds - Relationship: between the consumer and the brand - Reflection: it makes reference to the stereotypical user of the brand - Self-image: the way consumers see themselves when they buy or use a brand - Culture: system of values, source of inspiration and brand energy - Personality: reflects the set of human characteristics associated with a brand

de Chernatony (2010)	The distinctive or central idea of a brand and how the brand communicates this idea to different stakeholders	Five components: - Vision: provides a clear sense of direction - Culture: determinant factor to achieve the vision - Positioning: manifestation of the brand's functional values - Personality: it brings the brand's emotional values to life - Relationships with different stakeholders: staff, customers and other stakeholders

A strong brand identity has several benefits. Brand identity is an essential construct to gain sustainable competitive advantages, and effectively differentiate and manage brands (Aaker, 1996; Kapferer, 2004). Brand identity precedes and represents the basis for its image and reputation. A well managed brand identity may result in positive perceptions, attitudes and behaviours of different stakeholders. For instance, from the point of view of consumers, the creation of a unique, coherent and distinctive identity can add value to the company's products (Coleman et al., 2011) and increase their preference and loyalty (Johnson and Zinkhan, 1990; Simões et al., 2005). Likewise, brand identity can influence the attractiveness of the organisation as an employer. As such, brand identity can help companies to increase the motivation of the employees and attract better and more qualified applicants (Melewar, 2003), as well as greater investments (Arendt and Brettel, 2010; Van Riel, 1995).

Over the last decades, scholars and practitioners have proposed several brand identity frameworks (Aaker, 1996; Aaker and Joachisthaler, 2000; de Chernatony, 2010; Kapferer, 2004). Without being exhaustive, some of these frameworks are briefly described next (see Table 1). According to Aaker's framework (1996), one of the most referenced in the academic literature, brand identity consists of twelve dimensions organised around four perspectives: the brand as a product, the brand as a person, the brand as an organisation and the bran as a symbol. Kapferer (2004), on the other hand, offers an alternative proposal. Specifically, this author develops a brand identity prism with six interrelated facets. These six facets are: physique, relationship, reflection, self-image, culture and personality. Another interesting brand identity framework has been developed by de Chernatony (2010), who conceives brand identity in terms of five components: vision, culture, positioning, personality and relationships with different stakeholders.

Despite their interest, these brand identity frameworks are now without limitations. As such, although widely referenced in the academic literature, these proposals are conceptual in nature and have not been subject to empirical investigation (Coleman et al., 2011).

CONCEPTUAL FRAMEWORK AND HYPOTHESES

As mentioned before, most of the operationalisations and conceptual frameworks of corporate brand identity proposed in the literature are of a conceptual nature. Thus, in the last few years some researchers have tried to fill this gap by developing and validating measurement scales. The present research adopts the proposal developed by Coleman et al. (2011). After an extensive literature review, these authors conceptualise brand identity as a multidimensional construct comprised of five dimensions. The first dimension, employee and client focus, highlights the importance of understanding and attending the customers' needs, but also of appreciating the critical role played by employees within the organisation. Corporate visual identity is the second dimension. It refers to the set of visual cues such as the brand name, logo and slogan that make the organisation recognisable. These features also allow the transfer of meanings to the different stakeholders. The third dimension, brand personality, is a salient facet of different brand identity frameworks. It represents the set of human personality traits that are applicable to a brand. The fourth dimension refers to consistent communications, that is, the need for coordination and consistency in marketing communications. Finally, the fifth dimension, human resource initiatives, emphasises the central role played by employees as well as the training programmes designed for them.

The present research focuses on three relevant consequences of corporate brand identity management on employees' attitudes and behaviours. Specifically, the constructs explored are: employees' commitment to the organisation, employees' brand performance and cmployccs' satisfaction.

Commitment to the organisation is an important variable from a managerial perspective, given its influence on employees' attitudes and behaviours. Therefore, one of the central questions that guide this research is to analyse to what extent a well managed corporate brand identity can increase employees' commitment.

A lack of consensus exists on the definition of commitment. This is mainly due to the different theoretical approaches, as well as the numerous ways to conceptualise and measure this concept. From a general perspective, Meyer et al. (2006) define commitment as a force that binds an individual to a target (e.g. an organisation), and to a course of action of relevance to that target. Most definitions of this term consider it as a link or bond between the individual and the organisation (Mathieu and Zajac, 1990). As such, they mainly refer to the affective dimension of commitment, which is the focus of this study. This emotional attachment is considered as the main determinant of employees' attitudes and behaviours (Bloemer and Odekerken-Schröder, 2006).

Extant research on internal brand management posits that an appropriate management can increase employees' brand knowledge. This, in turn, can enhance the understanding of the brand strategies and decisions, as well as the values and promises associated to the brand (King and Grace, 2010). This mutual understanding of the brand, as a consequence of an internal brand management approach, can enhance the intellectual and emotional commitment to the organisation (Aurand et al., 2005; Thomson et al., 1999). In addition, employees who have a positive perception of the corporate brand identity and its management are more likely to develop a strong link or bond with the organisation, as well as a sense of pride and belonging to the company (Dukerich et al., 2002).

The different dimensions that underlie the brand identity construct can influence on employees' commitment to the organisation. For instance, Punjaisri and Wilson (2011) emphasise the positive effect that the initiatives related to the human resources or the communications have on employees' commitment. Other authors, such as Arendt and Brettel (2010), find that dimensions such as visual identity help employees internalise the brand, fostering the identification and commitment to it. Therefore, we can expect that in the search for greater trust and employee engagement, internal brand management and corporate brand identity are crucial elements to leverage employees' commitment (Burmann and Zeplin, 2005; Punjaisri and Wilson, 2011). Consequently, the following hypothesis is postulated:

Hypothesis 1

Corporate brand identity management has a positive effect on employees' commitment to the organisation.

The following hypotheses refer to the effect of employees' commitment on brand performance and employees' satisfaction.

Brand performance has been operationalised in a variety of ways, such as the brand market share or sales growth. In this research, this variable refers to the extent to which an employee performs his/her role, including the brand promise delivery, based on the brand standards (Punjaisri et al., 2009).

Results of previous research show a positive but moderate correlation between these two constructs (Riketta, 2002). Punjaisri and Wilson's (2011) work is one of the few studies that have empirically analysed the influence of commitment on brand performance. However, they could not find support for this relationship. Thus, more research is needed to shed additional light on this issue.

Contextual factors, such as the lack of access to some resources, and personal factors, such as employees' skills and abilities, could explain the weaknesses of this relationship. In this sense, these factors can reduce employee›s performance, thus preventing the effect of commitment on brand performance (Porter et al., 2005). Despite of this, according to previous research, it is expected that the greater the employees' commitment to the organisation, the greater the efforts made by employees to perform their functions and deliver the brand promise (de Chernatony and Segal-Horn, 2003). As such, it is traditionally assumed that employees who feel attached to their organisation and highly identify with their companies will have a higher performance. Committed employees are more likely to live the brand, reinforcing their sense of belonging the brand message (Boyd and Sutherland, 2006). Likewise, previous research has shown that high levels of affective commitment result in numerous benefits related to employees' performance. For instance, employees' commitment can increase the willingness to develop behaviours that benefit to the organisation (Strauss et al., 2009). Similarly commitment positively relates with service quality and service recovery performance (Malhotra and Mukherjee, 2004). Therefore, it is hypothised that:

Hypothesis 2

Employees' commitment to the organisation has a positive effect on their brand performance.

The last relationship analysed in the model refers to the influence of employees' commitment on their satisfaction. Job satisfaction is defined as "the pleasurable emotional state resulting from the appraisal of one's job as achieving or facilitating the achievement of one's job values" (Locke, 1969, p. 316). Its measurement and analysis is particularly interesting, as it can serve as a diagnostic tool to explore the situation of the employees in the organisation and identify problems or causes of dissatisfaction.

From the field of organisational psychology, it is postulated that the links and bonds developed by employees towards the organisation influence the way they think, feel and behave in the workplace (Ashforth and Mael, 1989; Haslam et al., 2003). In this sense, several works have found a positive relationship between employees commitment to the organisation and their level of satisfaction (Jones et al., 2003; King and Grace, 2010).

When employees identify themselves with an organisation, and are committed to the company, they tend to feel better about both themselves and their work. Consequently, it is more likely that committed employees feel more satisfied and motivated to perform their work (Harris and de Chernatony, 2001; Wheeler et al., 2006). Therefore, the following hypothesis is proposed:

Hypothesis 3

Employees' commitment to the organisation has a positive effect on their satisfaction.

Fig. 1 presents the relationships proposed in this study. As can be seen in the figure, corporate brand identity is a multidimensional construct comprised of five dimensions influencing employees' attitudes and behaviours.

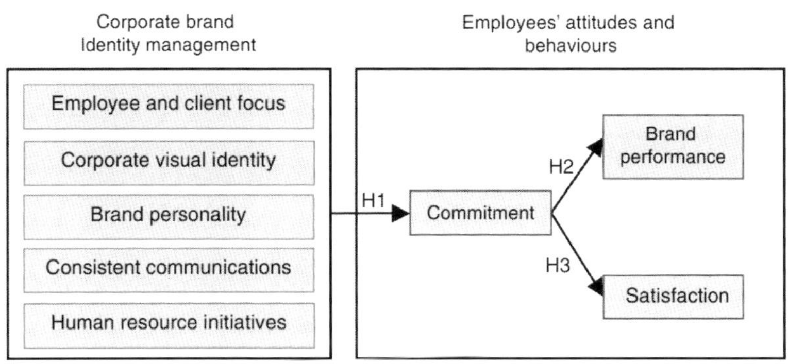

Figure 1: Conceptual model.

METHOD

Sample and procedure

To test the hypotheses, a study was undertaken during January and February 2013 with employees of the main banks in the United Kingdom. Data was collected by an international market research company using an online questionnaire. In total, 297 valid questionnaires were obtained.

Males represent 46.8% of the sample and females 53.2%. 7.1% of respondents are between 18 and 24 years old; 35% between 25 and 24 years old; 34% between 35 and 44 years old, and 23.9% between 44 and 67 years old. Finally, in terms of employment history, 41% of respondents have been working in the bank between 1 and 5 years, 21% 6–10 years, 20% 11–20 years, and 18% more than 20 years.

Measures

A literature review was undertaken to select the most appropriate way to measure each variable. In all cases, 7-point Likert-type questions were used with 1 = strongly disagree and 7 = strongly agree. Items proposed by Coleman et al. (2011) were used to measure corporate brand identity. According to these authors, brand identity is a multidimensional construct comprised of five dimensions: employee and client focus, visual identity, brand personality, consistent communications, and human resource initiatives. Organisational commitment was measured using the items proposed by King and Grace (2010). Brand performance was assessed following Punjaisri and Wilson (2011). Finally, based on King and Grace (2010), four items were used to measure employee satisfaction. A list of the items used to measure each construct is provided in Table 2.

Table 2: Scales items included in the questionnaire.

Brand corporate identity management (Coleman et al., 2011)
Employee and client focus
ECF1. Our top management is committed to providing quality service
ECF2. Our bank treats every employee as an essential part of the organisation
ECF 3. Our employees will help clients in a responsive manner
ECF 4. Our bank makes an effort to discover our clients, needs
ECF 5. Our bank responds to our clients' needs
Corporate visual identity
CVI1. The corporate visual identity is helpful in making our bank recognisable
CVI2. The font we use is an important part of our visual identity
CVI3. Our logo is an important part of who we are
Brand personality
BP1. The associations making up our brand personality are extremely positive
BP2. Our clients have no difficulty describing our brand personality
BP3. Our brand personality has favourable associations

Consistent communications
CC1. The people managing the communications programme for our bank have a good understanding of the strengths and weaknesses of all major marketing communications tools
CC2. Our bank's advertising, PR and sales promotion all present the same clear consistent message to our stakeholders
Human resource initiatives
HRI1. Our employee training programmes are designed to develop skills required for acquiring and deepening client relationships
HRI2. Our bank regularly monitors employees' performance
Commitment (King and Grace, 2010)
COM1. I am proud to be part of this bank
COM2. I really care about the fate of this bank
COM3. I feel like I really fit in this bank
Brand performance (Punjaisri and Wilson, 2011)
PER1. I can successfully fulfil responsibilities specified in my job descriptions
PER2. The quality of the work I do matches brand standards of my bank
PER3. I deliver the promise that my bank's brand has with customers
PER4. I always deliver the required level of service in response to customers' specific requests
Satisfaction (King and Grace, 2010)
SAT1. I feel reasonably satisfied with my job
SAT2. I feel a great sense of satisfaction from my job
SAT3. I am satisfied with my overall job

ANALYSES AND RESULTS

Hypotheses were tested using structural equation modelling (SEM). The data was analysed using EQS 6.2 and the robust maximum-likelihood estimation. First, the psychometric properties of the scales were examined. Then, the structural model was evaluated by testing the hypotheses.

Table 3 presents the descriptive statistics for all the variables included in the study.

Table 3: Descriptive statistics and correlations.

	Mean	S.D.	1	2	3	4	5	6	7	8
1. Employee and client focus	5.37	1.34	0.764							
2. Corporate visual identity	5.58	1.24	0.496	0.697						
3. Brand personality	5.22	1.33	0.587	0.498	0.755					
4. Consistent communications	5.25	1.30	0.498	0.352	0.542	0.795				
5. Human resource initiatives	5.16	1.60	0.436	0.220	0.440	0.537	0.850			
6. Commitment	5.12	1.62	0.504	0.238	0.473	0.462	0.496	0.850		
7. Brand performance	5.93	1.06	0.394	0.375	0.314	0.397	0.303	0.354	0.744	
8. Satisfaction	5.00	1.69	0.342	0.179	0.379	0.334	0.429	0.776	0.338	0.909

Note: Means and standard deviations are based on summated scale averages. Items deleted in the validation process are not included. Squared correlations are below the diagonal and AVE estimates are presented on the diagonal.

Scales validation

To assess the initial reliability of the measures, Cronbach's alpha and the item-total correlation for all scales were used. Cronbach's alpha for all the constructs was above 0.70 (Nunnally, 1978). Furthermore, the item-to-total correlations were all above the threshold of 0.30 (Nurosis, 1993). There was only one exception. The Cronbach's alpha for the human resource initiatives scale (alpha = 0.672). We then performed exploratory factor analyses using principal components analysis with varimax rotation. Results suggested that the corresponding items of each scale clustered into a single factor with significant factor loadings. The explained variance exceeded 60% in each case.

To attempt to identify potential problems with common method variance, basing on the procedure recommended by Podsakoff et al. (2003), Harman's one-factor test was used. The results of the factor analyses revealed that more than a single factor emerged with no single factor accounting for the majority of variance. Thus, there is no evidence to suggest the presence of common method bias.

Scales were then evaluated using confirmatory techniques to assess reliability, dimensionality and validity. Results suggested the deletion of item HR2, as its R^2 was below 0.5. After this deletion, fit indices were above the suggested 0.90 cutoff point, thus providing evidence that the final measurement model fit the data appropriately. Further, the RMSEA index was below the 0.05 cutoff point (Hair et al., 2006) (see Table 4).

Table 4: Confirmatory factor analysis results.

Factor	**Items**	λ	R^2	**CR**	**AVE**
Employee and client focus	ECF1.	0.800	0.641	0.941	0.764
	ECF2.	0.779	0.606		
	ECF3.	0.896	0.803		
	ECF4.	0.939	0.882		
	ECF5.	0.942	0.887		
Corporate visual identity	CVI1.	0.886	0.785	0.873	0.697
	CVI2.	0.770	0.593		
	CVI3.	0.845	0.714		
Brand personality	BP1.	0.882	0.777	0.902	0.755
	BP2.	0.868	0.753		
	BP3.	0.856	0.732		

Consistent communications	CC1.	0.861	0.741	0.886	0.795
	CC2.	0.921	0.848		
Human resource initiatives	HRI1.*	0.922	0.850	0.850	0.850
Commitment	COM1.	0.930	0.865	0.944	0.850
	COM2.	0.886	0.785		
	COM3.	0.949	0.901		
Brand performance	PER1.	0.835	0.697	0.921	0.744
	PER2.	0.905	0.819		
	PER3.	0.896	0.803		
	PER4.	0.811	0.658		
Satisfaction	SAT1.	0.946	0.895	0.968	0.909
	SAT2.	0.956	0.915		
	SAT3.	0.958	0.917		
Fit indices					
S-Bχ^2 (df, p) = 360.78 (225, $p < 0.001$); RMSEA = 0.045 NNFI = 0.942; CFI = 0.953; IFI = 0.954					

Note: λ, standardised factor loadings; R^2, regression coefficients; CR, composite reliability; AVE, average variance extracted; S-B χ^2, Satorra–Bentler scaled chi-square; RMSEA, root mean square error of approximation; NNFI, non-normed fit index; CFI, comparative fit index; IFI, incremental fit index. *Reliability for the single item measure was fixed at 85% (Jöreskog and Sörbom, 1993).

All factor loadings were above 0.5 and statistically significant which suggested the convergent validity of the scales (Steenkamp and Van Trijp, 1991). Likewise, the coefficients had a clear relation with their underlying factors ($R^2 > 0.5$). Constructs were shown to possess high internal validity.

As such, the average variance extracted (AVE) and composite reliability (CR) values were greater than 0.5 and 0.7 respectively (Hair et al., 2006).

Discriminant validity was also supported. As such, none of the confidence intervals around the correlation estimate between any two factors contained a value of one (Anderson and Gerbing, 1988). In addition, as can be seen in Table 3, the AVE for any two constructs was always greater than the square of the correlation estimate (Fornell and Larcker, 1981).

Structural model

Once the psychometric properties of the scales were examined, the structural model was tested. As a multidimensional construct, first the average of the items within each of the five corporate brand identity dimensions was calculated. These averaged items were then used as manifest indicators of the global corporate brand identity construct (Baldauf et al., 2009; Santos and Álvarez, 2008).

The analysis yielded a good overall fit, as shown in Fig. 2. As proposed in Hypothesis 1, corporate brand identity management proved a significant predictor for employees' commitment (β = 0.822; *t*-value = 15.525). Therefore, employees' commitment will increase when organisations focus their efforts on the management of its corporate brand identity.

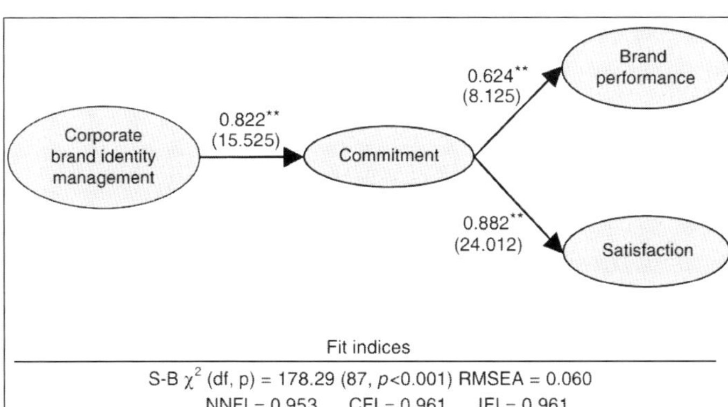

Figure 2: Structural model results.

Note: ** *P*<0.05

As can be seen in Fig. 2, Hypothesis 2, testing the relationship between employees' commitment and brand performance, received statistical support (β = 0.624; *t*-value = 8.125). Therefore Hypothesis 2 was supported, indicating that employees' commitment will favour employees' performance on their jobs and tasks.

In addition, results show that employees' commitment was positively associated with their level of satisfaction on the job. As such, commitment had a positive and significant effect on satisfaction (β = 0.882; *t*-value = 24.012). Therefore, the last hypothesis was also supported.

Finally, results reveal that corporate brand identity management had a significant indirect positive effect on employees' brand performance ($z = 7.23, p < 0.001$) as well as on the level of employees' satisfaction on the job ($z = 12.98, p < 0.001$), according to Sobel›s test results.

CONCLUSIONS, IMPLICATIONS AND DIRECTIONS FOR FURTHER RESEARCH

The purpose of the present study was to gain a deeper understanding of the corporate brand identity concept by analysing the effect that its management has on employees' attitudes and behaviours. The findings confirm the relevance of this construct, given its ability to increase employees' commitment, which in turn influences their brand performance and satisfaction with their jobs.

Following the proposal of Coleman et al. (2011), corporate brand identity management is conceptualised as a multidimensional construct comprised by five dimensions: employee and client focus, corporate visual identity, brand personality, consistent communications and human resource initiatives.

Findings suggest that corporate brand identity management encourages the commitment of the employees to the organisation. Previous works have highlighted the importance of the degree of employees' commitment as a key driver of success and differentiation for both the brand and the organisation itself (Balmer, 2001). A high level of commitment can be achieved through different ways. However, the strength of the causal relationship found in this study confirms that corporate brand identity management is an important antecedent to create this bond between the employee and the organisation.

The research also reveals some interesting findings regarding the relationship between employees' commitment and brand performance and employee satisfaction. First, contrary to other studies, which found a weak or nonexistent relationship between employees' commitment and brand performance (e.g. Punjaisri and Wilson, 2011), this research corroborates that commitment influences the way an employee performs his/her role based on the brand standards. In addition, the results indicate that commitment positively affects employee satisfaction. Based on these findings, it is possible to state that managers should foster employees' commitment to the organisation.

Finally, it should be noted that corporate brand identity management has an indirect effect on both brand performance and employee satisfaction, through a higher employees' commitment.

These findings have implications for both scholars and managers. From a theoretical perspective, the present study enriches corporate brand identity research addressing some of the limitations regarding other brand identity studies. As such, from the employees' point of view, it adopts an empirical perspective, which complements the theoretical works developed to date. As earlier explained, this fact is specially relevant since for a long time the role of the employees has been largely ignored, despite their importance in the creation and communication of the corporate brand identity.

The research findings have also several managerial implications. First, we highlight the relevance of the corporate brand identity to the organisation, as well as the operationalisation of this construct. Traditionally, brand image management has been the main concern of many managers. The interdependent relationship between both brand image and brand identity, and the benefits derived from a suitable identity management, advise, however, that managers take into account this relevant concept. This recommendation is even more important in the financial services sector. As such, given the low level of differentiation among financial products and services (Wilkinson and Balmer, 1996), and the global economic crisis, which has contributed to the creation of a climate of uncertainty and distrust, it is needed to emphasise the importance of a well brand identity management (de Chernatony and Cottam, 2006).

The creation and management of the corporate brand identity requires, however, understanding the dimensions that underlie this construct. In many cases, corporate brand identity is exclusively associated with some of its dimensions, such as the visual identity, ignoring its complexity and multiple facets. This research provides a guide for managers seeking to identify the key dimensions in which they should focus. Findings suggest that not only tangible aspects of brand identity, such as visual identity, or dimensions that are most directly related to the employees, such as the human resources initiatives, are relevant. By contrast, it is also important that organisations pay attention to both the customers' and employees' needs. In addition, it is essential to create training programmes designed to develop the skills required for deepening client relationships. Furthermore, it is necessary to be aware of the fact that a brand or an organisation without personality lack of identity. Therefore, organisations should focus on those favourable associations and attributes, which will let them differentiate from others. Finally, it is very important that the organisation sets up a coherent and controlled process of its corporate communication. This process will imply the definition of the specific marketing communications tools to be used (e.g. advertising, public relations, etc.) and the kind of information related to the brand and the company that will be communicated.

Employees' attitudes and behaviours play a key role in achieving a high level of customer satisfaction. Accordingly, banks and organisations in general, should pay special attention to the facets that are part of the corporate brand identity (employee and client focus, corporate visual identity, brand personality, consistent communications and human resource initiatives). This will foster the bonds between the employees and the organisation, and as a consequence, their performance and satisfaction.

There are several limitations to this study that suggest directions for further research. First, the study was conducted in the UK banking sector. This sector has special characteristics not shared by other organisations in others sectors. Thus, the generalisability of the results is limited to this context. Further research should, therefore, consider the applicability of findings in other sectors, countries and cultures.

Furthermore, future research could examine additional antecedents of corporate identity brand management, such as the level of brand orientation within the organisation (Urde et al., 2013), or other variables of interest, such as organisational values. Likewise, future studies could include additional outcomes related to the attitudes and behaviours of the employees and other stakeholders. In this sense, a possible extension would be to adopt a dyadic perspective that includes both employees and customers. Undertaking this research could be helpful to reach a better understanding of the corporate brand identity management and its consequences.

Finally, the present study used cross-sectional data, that is, data were obtained at one point in time. Consequently, a longitudinal framework would provide more insight into probable causation.

Despite these limitations, this study is a step towards a fuller understanding of the importance of the corporate brand identity management and its effects on employees' commitment, brand performance and satisfaction. It is hoped that the findings presented in this research guide managers' decisions about this important intangible asset.

ACKNOWLEDGEMENTS

The authors acknowledge the financial support of the I + D + I projects (Ref: ECO2009-08283; ECO2013-41257-P) from the Government of Spain and the project "GENERES" (Ref: S-09) from the Government of Aragon and the European Social Fund. Isabel Buil would also like to thank the financial support received from the Caja de Ahorros de la Inmaculada of Zaragoza and the DGA (CONSI + D) (Programa Europa XXI de Estancias de Investigación). Eva Martínez also thanks the financial support of the "Subprograma de estancias de movilidad de profesores e investigadores séniores en centros extranjeros de enseñanza superior e investigación, incluido el Programa Salvador de Madariaga" (Ref: PR2011-0352).

REFERENCES

1. Aaker, A., 1996. Building Strong Brands. The Free Press, New York.
2. Aaker, A., Joachisthaler, E., 2000. Brand Leadership. The Free Press, New York.
3. Alsem, K.J., Kostelijk, E., 2008. Identity based marketing: a new

balanced marketing paradigm. Eur. J. Market. 42 (9/10), 907---914.

4. Anderson, J.C., Gerbing, D.W., 1988. Structural equation modelling in practice: a review and recommended two step approach. Psychol. Bull. 103, 411---423.
5. Arendt, S., Brettel, M., 2010. Understanding the influence of corporatc social rcsponsibility on corporate identity, image, and firm performance. Manag. Decis. 48 (10), 1469---1492.
6. Ashforth, B.E., Mael, F., 1989. Social identity and the organization. Acad. Manag. Rev. 14 (January), 20---39.
7. Aurand, T.W., Gorchels, L., Bishop, T.R., 2005. Human resource management's role in internal branding: an opportunity for cross-functional brand message synergy. J. Prod. Brand Manag. 14 (2), 163---169.
8. Baldauf, A., Cravens, K.S., Diamantopoulos, A., Zeugner-Roth, K.P., 2009. The impact of product-country image and marketing efforts on retailer-perceived brand equity: an empirical analysis. J. Retail. 85 (4), 437---452.
9. Balmer, J.M., 2001. Corporate identity, corporate branding and corporate marketing --- seeing through the fog. Eur. J. Market. 35 (3/4), 248---291.
10. Balmer, J.M., 2008. Identity based views of the corporation: insights from corporate identity, organisational identity, social identity, visual identity, corporate brand identity and corporate image. Eur. J. Market. 42 (9), 879---906.
11. Balmer, J.M., 2012. Strategic corporate brand alignment: perspective from identity based views of corporate brands. Eur. J. Market. 46 (7/8), 1064---1092.
12. Balmer, J.M., Greyser, S.A., 2002. Managing the multiple identities of the corporation. Calif. Manag. Rev. 44 (3), 72---86.
13. Bloemer, J., Odekerken-Schröder, G., 2006. The role of employee relationship proneness in creating employee loyalty. Int. J. Bank Market. 24 (4), 252---264.
14. Blombäck, A., Ramírez-Pasillas, M., 2012. Exploring the logics of corporate brand identity formation. Eur. J. Market. 17 (1), 7---28.

15. Boyd, G., Sutherland, M., 2006. Obtaining employee commitment to living the brand of the organization. S. Afr. J. Bus. Manag. 37 (1), 9---20.
16. Burmann, C., Zeplin, S., 2005. Building brand commitment: a behavioural approach to internal brand building. J. Brand Manag. 12 (4), 279---300.
17. Burmann, C., Zeplin, S., Riley, N., 2009. Key determinants of internal brand management success: an exploratory empirical analysis. J. Brand Manag. 16 (4), 264---284.
18. Coleman, D., de Chernatony, L., Christodoulides, G., 2011. B2B service brand identity: scale development and validation. Ind. Market. Manag. 40 (7), 1063---1071.
19. Colton, S., Oliveira, P., 2009. Banking On It. The Role of the Corporate Brand in Rebuilding Trust. Interbrand: Creating and Managing Brand Value, Available at: http://www.interbrand.com
20. Currás, R., 2010. Identidad e imagen corporativas: revisión conceptual e interrelación. Teoría y Praxis 7, 9---34.
21. da Silveira, C., Lages, C., Simoes, C., 2013. Reconceptualizing brand identity in a dynamic environment. J. Bus. Res. 66, 28---36.
22. de Chernatony, L., 2010. From Brand Vision to Brand Evaluation, 3rd ed. Butterworth-Heinemann, Oxford.
23. de Chernatony, L., Cottam, S., 2006. Why are all financial services brands not great? J. Prod. Brand Manag. 15 (2), 88---97.
24. de Chernatony, L., Segal-Horn, S., 2003. The criteria for successful services brands. Eur. J. Market. 37 (7/8), 1095---1118.
25. Dukerich, J.M., Golden, B.R., Shortell, S.M., 2002. Beauty is in the eye of the beholder: the impact of organizational identification, identity, and image on the cooperative behaviors of physicians. Adm. Sci. Q. 47 (3), 507---533.
26. Fornell, C., Larcker, D.F., 1981. Evaluating structural equation models with unobservable variables and measurement error. J. Market. Res. 18 (1), 39---50.
27. Hair, J.F., Black, B., Babin, B., Anderson, R.E., Tatham, R.L., 2006. Multivariate Data Analysis, 6th ed. Prentice Hall, Upper Saddle River, NJ.

28. Harris, F., de Chernatony, L., 2001. Corporate branding and corporate brand performance. Eur. J. Market. 35 (3/4), 441---456.
29. Haslam, A., Postmes, T., Ellemers, N., 2003. More than a metaphor: organizational identity makes organizational life possible. Br. J. Manag. 14 (4), 357---369.
30. Jones, E., Busch, P., Dacin, P., 2003. Firm market orientation and salesperson customer orientation: interpersonal and intrapersonal influences on customer service and retention in business-to-business buyer---seller relationships. J. Bus. Res. 56, 323---340.
31. Johnson, M., Zinkhan, G.M.,1990. Defining and measuring company image. In: Thirteenth Annual Conference of the Academy of Marketing Science. Academy of Marketing Science, New Orleans, LA.
32. Jöreskog, K.G., Sörbom, D., 1993. LISREL 8 User's Guide. Scientific Software International, Chicago.
33. Kapferer, J.N., 2004. The New Strategic Brand Management. Creating and Sustaining Brand Equity Long Term. Kogan Page, London.
34. King, C., Grace, D., 2010. Building and measuring employee-based brand equity. Eur. J. Market. 44 (7), 938---971.
35. Locke, E.A., 1969. What is job satisfaction? Organ. Behav. Hum. Perform. 4 (4), 309---336.
36. Malhotra, N., Mukherjee, A., 2004. The relative influence of organisational commitment and job satisfaction on service quality of customer-contact employees in banking call centres. J. Serv. Market. 18 (3), 162---174.
37. Mathieu, J.E., Zajac, D.M., 1990. A review and meta-analysis of the antecedents, correlates, and consequences of organizational commitment. Psychol. Bull. 108 (2), 171.
38. McEwen, W.J., 2010. Which Bankers Deserve the Bonus, Available at: http://gmj.gallup.com/content/139748/bankers-deservebonus.aspx
39. Melewar, T.C., 2003. Determinants of the corporate identity construct: a review of the literature. J. Market. Commun. 9 (4), 195-

--220.

40. Meyer, J.P., Becker, T.E., Van Dick, R., 2006. Social identities and commitments at work: toward an integrative model. J. Org. Behav. 27 (5), 665---683.
41. Nunnally, J.C., 1978. Psychometric Theory, 2nd ed. McGraw-Hill, New York.
42. Nurosis, M.J., 1993. SPSS Statistical Data Analysis. SPSS Inc.
43. Podsakoff, P.M., MacKenzie, S.B., Lee, J.Y., Podsakoff, N.P., 2003. Common method biases in behavioral research: a critical review of the literature and recommended remedies. J. Appl. Psychol. 88 (5), 879---903.
44. Porter, L.W., Steers, R.M., Mowday, R.T., 2005. Do Employee Attitudes Towards Organizations Matter? The Study of Employee Commitment to Organizations. Great Minds in Management. Oxford University, New York.
45. Punjaisri, K., Wilson, A., 2011. Internal branding process: key mechanisms, outcomes and moderating factors. Eur. J. Market. 45 (9/10), 1521---1537.
46. Punjaisri, K., Wilson, A., Evanschitzky, H., 2009. Internal branding to influence employees' brand promise delivery: a case study in Thailand. J. Serv. Manag. 20 (5), 561---579.
47. Riketta, M., 2002. Attitudinal organizational commitment and job performance: a meta-analysis. J. Org. Behav. 23 (3), 257---266.
48. Santos, M.L., Álvarez, L.I., 2008. Efectos de la gestión de calidad total en la transformación en la innovación tecnológica y administrativa. Cuadernos de Dirección y Economía de la Empresa 37, 33---66.
49. Simões, C., Dibb, S., Fisk, R.P., 2005. Managing corporate identity: an internal perspective. J. Acad. Market. Sci. 33 (2), 153---168.
50. Steenkamp, J.-B.E.M., Van Trijp, H.C., 1991. The use of LISREL in the validating marketing constructs. Int. J. Res. Market. 8, 283---299.
51. Strauss, K., Griffin, M.A., Rafferty, A., 2009. Proactivity directed toward the team and organization: the role of leadership commitment and role-breadth self-efficacy. Br. J. Manag. 20, 279---291.

52. Thomson, K., de Chernatony, L., Arganbright, L., Khan, S., 1999. The buy-in benchmark: how staff understanding and commitment impact brand and business performance. J. Market. Manag. 15 (8), 819---835.
53. Urde, M., Baumgarth, C., Merrilees, B., 2013. Brand orientation and market orientation --- from alternatives to synergy. J. Bus. Res. 66 (1), 13---20.
54. Van Riel, C., 1995. Principles of Corporate Communication. Prentice-Hall, London, UK.
55. Wallace, E., de Chernatony, L., Buil, I., 2013. Building bank brands: how leadership behavior influences employee commitment. J. Bus. Res. 66 (2), 165---171.
56. Wheeler, A.R., Richey, R.G., Tokkman, M., Sablynski, C.J., 2006. Retaining employees for service competency: the role of corporate brand identity. J. Brand Manag. 14 (1), 96---113.
57. Wilkinson, A., Balmer, J.M., 1996. Corporate and generic identities: lessons from the Co-operative Bank. Int. J. Bank Market. 14 (4), 22---35.

CHAPTER 10

THE EVOLUTION OF STRATEGIC MANAGEMENT RESEARCH: RECENT TRENDS AND CURRENT DIRECTIONS

Luis Ángel Guerras-Martín[a], Anoop Madhok [b], Ángeles Montoro-Sánchez[c]

[a] Universidad Rey Juan Carlos, Spain

[b] Schulich School of Business, York University, Canada

[c] Universidad Complutense de Madrid, Spain

ABSTRACT

Strategic management is a relatively youthful discipline that has steadily matured over the past fifty years. The field has become consolidated over this period, while simultaneously expanding the range of topics analyzed and research methodologies used. Different theories and approaches, addressing different research topics, have been developed to explain the reasons underlying firms' competitive advantage and success. In this paper, we posit the existence of two pendulums in constant motion that, on the one hand, reflect the tension that has historically existed between the focus on internal firm factors and external environmental attributes respectively and, on the other hand, the tension between a more macro level of analysis, i.e., the firm and its environment, and a more micro level one, i.e., individuals and their relations within the firm. The frontier of research in strategic management is shaped by the simultaneous movement of both pendulums.

Keywords: Strategic management, Approaches to research, Evolution and trends

INTRODUCTION: TRENDS IN STRATEGIC MANAGEMENT RESEARCH

Strategic management is a youthful discipline. Its origins date back to the 1960s, with its roots to be found mainly in the seminal publications by Chandler (1962), Ansoff (1965) and Andrews (1971). Since then, it has evolved significantly, becoming an ever more mature and consolidated field within the realm of management.

The discipline›s progress toward maturity has been accompanied by several factors. First, there has been a marked increase in the range of topics addressed (Hoskisson et al., 1999). The study of "best practices" in the 1960s has given way to an analysis of such varied topics as internationalization, cooperation between firms, strategies and competition in the markets for products and factors, strategic leadership, and the relationship between a firm›s strategy and its corporate social responsibility, to mention just a few.

Second, there has been significant growth in the range of research methods used, with these becoming steadily more sophisticated (Hoskisson et al., 1999; Ketchen et al., 2008). In-depth case studies have been largely replaced by the use of quantitative tools based on complex econometric techniques, multilevel analysis and, more recently, hybrid methodologies, whereby a single study combines quantitative and qualitative techniques, with each being adapted to the nature of the problem to be analyzed (Molina-Azorín, 2012).

Third, a further indicator of the discipline›s growing maturity is the increasing consensus regarding certain basic notions, such as the definition or concept of strategy or strategic management. Ronda-Pupo and Guerras-Martin (2012) reveal how consensus around the concept of strategy has been building up and spreading over time in a slow but inexorable manner. Although the diversity of definitions continues to be a common feature of our discipline, more aspects of the concept have become part of the core of the definition over the years or drawn closer to it. It is both surprising and noteworthy to discover how major aspects of the concepts of strategy and strategic management are so similar in the two studies conducted in different contexts and with differing methodologies. Using implicit and explicit definitions from a set of scholars, Nag et al. (2007) identify seven key components of the concept of strategic management: performance, firms, strategic initiatives, environment, internal organization, managers/owners and resources. Although they employed a different approach based on co-word analysis, the same components feature among the more relevant items underlying the concept of strategy in the research by Ronda-Pupo and Guerras-Martin (2012).

Fourth, and finally, the academic community interested in strategy research has been growing steadily, not just in terms of the number of scholars dedicated to this field but also their international nature as well as the linkages among them. If we take our yardstick to be the publication *Strategic Management Journal* (SMJ), we can see how cooperation has been increasing between scholars from both an inter-institutional as well as an international perspective (Ronda-Pupo and Guerras-Martin, 2010; Guerras-Martín and Ronda-Pupo, 2013). Furthermore, the Strategic Management Society currently has almost 3000 members from more than 80 countries.

Over the course of its development, strategic management has steadily accumulated a reasonably large corpus of knowledge, both of a theoretical and an empirical and methodological nature. Accordingly, there has been a need every so often to define the discipline's state-of-the-art and its direction at that moment. This has been reflected in three different ways. First, there are numerous works that have reflected upon the discipline itself, through books, papers, dedicated issues of journals or specific handbooks on the subject. As examples, we may mention the books by Fredrickson (1990), Rumelt et al. (1994), and Pettigrew et al. (2002a).

Second, there has been an upsurge of dedicated issues of journals that have sought to explore the frontiers of knowledge in the field and its links to other disciplines. The journal SMJ, for example, has published special issues on such relevant topics as networks and alliances, organizational capabilities, entrepreneurship, global strategy, strategic process, resource-based view (RBV), evolutionary approaches, technological competences, strategy and economics and the psychological foundations of strategic management, among others (Guerras-Martín and Ronda-Pupo, 2013).

Third, recent years have witnessed the emergence of a line of research that uses scientometric techniques to discover and analyze the intellectual structure of strategic management and its evolution. We thus encounter research that analyzes the concept of strategy or strategic management (Nag et al., 2007; Ronda-Pupo and Guerras-Martin, 2012), the more salient lines and topics of research (Furrer et al., 2008), the intellectual structure and the more influential papers and authors (Ramos-Rodríguez and Ruiz-Navarro, 2004; Nerur et al., 2008; Furrer et al., 2008), the foremost journals and their profiles (Azar and Brock, 2008; García-Merino and Santos-Álvarez, 2009), or the structure and development of the international academic community (Ronda-Pupo and Guerras-Martin, 2010).

These instances of research that use scientometric tools do not replace the works that review and reflect upon the discipline, but instead complement them appropriately. By introducing quantitative techniques, they allow identifying and measuring specific variables related to the discipline and its development, thereby facilitating an objective approach to the analysis of such variables. Moreover, they permit the discovery of relationships between researchers and topics that are not obvious at first glance, such as the intellectual structure or the key terms that encapsulate myriad definitions of a concept.

Based on the above, our objective in this paper is to analyze past and current trends in strategic management research, a field characterized by assorted and manifold aspects of relevance. We pull together the various and diverse theoretical perspectives informing this research through utilizing the image of a double pendulum swinging in unison. This involves identifying and analyzing the tension between the internal and external domains of strategic management research as well as the tension between the macro and micro levels of analysis. Following this, we present the works included in this monograph, which seek to describe and explore some of the research frontiers.

THE EVOLUTION OF STRATEGIC MANAGEMENT RESEARCH: THE DUAL PENDULUM

Strategic management has been regarded as a fundamental issue that explains the success or failure of firms (Rumelt et al., 1994). This entails discovering why certain firms are successful while others are not; in other words, identifying the factors of success. Although this core motivation is shared by all the researchers in this field, the path to be pursued is not quite so clear. This is so because, among other reasons, the eclectic and multidisciplinary nature of strategic management (Hoskisson et al., 1999) attracts the interests of researchers from different disciplines: economics, organizational theory, sociology, psychology, management, etc. As a result, researchers often have different backgrounds, approaches or foci of interest. On top of this, since research into strategy is linked to firm operations, many tools that are now considered the staples of any analysis have emerged from the firms themselves or from the strategic consultancies that have advised them (Pettigrew et al., 2002b).

This situation has meant that the quest to find the factors of success has focused on different firm aspects and variables of interest. Mintzberg (1990) reminds us of the ancient fable retold in verse by John Godfrey Saxe (19th century US poet), in which each one of the blind men (researchers) sets out to describe the elephant (discipline) solely through touch (approach), whereby each one describes a different part of the animal. They are all correct in their description, yet it is difficult to deduce what the elephant actually looks like from each one of the individual descriptions.

One way of making sense of this diversity of factors is to classify them according to some criterion. Broadly, on the one hand many researchers have tended to focus their attention on a firm›s internal factors, such as its strengths and weaknesses. On the other hand, the reasons for success have been sought in the environment›s opportunities and threats. An alternative and complementary criterion enables us to organize the research according to the level of analysis adopted. This may involve the firm as a whole – macro level – or specific aspects of it linked to the behavior of its individuals – micro level.

In line with this, one can identify two types of tensions throughout the development of the strategic management field: one between internal and external considerations and the other between macro and micro level considerations. Typically, certain periods have recorded a prevalence of one or the other aspect in the research. This situation can be likened to a dual pendulum, following the metaphor used by Hoskisson et al. (1999), where the focal thrust is on internal or external considerations or on macro or micro level issues. The two pendulums have been moving simultaneously over the course of the discipline›s history, and their overall movements have defined the discipline›s evolution and its current state. Fig. 1 shows the arrangement of the main approaches and theories involved in strategic management according to these two classification criteria.

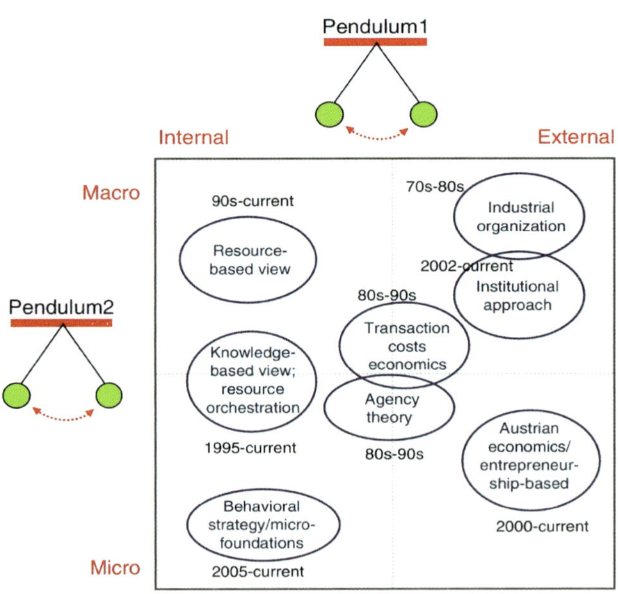

Figure 1: Evolution of strategic management research.

THE TENSION BETWEEN A FIRM'S INTERIOR AND ITS ENVIRONMENT

For many scholars reflecting upon the discipline, one of the key issues underlying its evolution is the primacy of attention given to internal or external considerations when explaining firm success. Hoskisson et al. (1999) popularized the metaphor of the pendulum and its swings to explain how strategic management has always tended to look for the factors of success either inside or outside a firm. For these authors, the pendulum started in the 1960s, when the focus was on analyzing internal strengths and weaknesses and the aim, therefore, was to look inside successful firms for those factors that underlay and had driven their performance.

In our opinion however, it is not so clear that the attention of the so-called founders of strategy was focused solely on internal considerations. We would contend instead that, during the 1960s and 1970s, more general approaches or frameworks were considered that laid the foundations for subsequent theories. Thus, strategy was seen as the way to link a firm to the environment in which both the internal and external aspects were important. This was mirrored in the SWOT analysis that investigated both the interior of firms (strengths and weaknesses) and their exterior (opportunities and threats), or in the strategic matrices (BCG growth-share matrix, or McKinsey market attractiveness-business strength) which merged both aspects – firm and industry environment – in a single analytic tool.

Toward the end of the 1970s and in the 1980s, the search for the keys to success gravitated toward the external environment, with research on the industry structure becoming paramount (Porter, 1980). The economics of organizations, through the contributions made by agency theory (Jensen and Meckling, 1976) and transactions cost theory (Williamson, 1975, 1985), shifted the pendulum toward more of a middle position that addressed both the internal and external aspects in the search for success. Finally, the appearance and rise of the resource-based view (RBV), stimulated by the works of Wernerfelt (1984) and Barney (1991), once again swung the pendulum into the interior of firms, to some extent going back to the beginning.

Although the pendulum metaphor is highly illustrative of the development of strategic management, with its focus on certain groups of factors that explain the success of firms, its strict application would be somewhat simplistic. It is one thing to reduce a research analysis to internal or external factors, for reasons of purpose or methodology, and it is another and very different matter to assume that the reality consists solely of one or the other. Clearly, both aspects are important. For instance, although it is important to choose an appropriate industry, with the industry›s appeal being undoubtedly an explanatory factor of success, yet the availability of valuable resources that set a firm apart from its competitors and which it can use to build a sustainable competitive advantage is no less important.

Although it is true that research in strategy has paid more attention in recent years to internal rather than external factors, the latter have not in any way disappeared from priority lines of research. Furrer et al. (2008) show how, over 26 years of research into strategy (1980–2005), the research stream linked to resources has been growing on a slow but steady basis. Whereas over the 1980–1985 period it accounted for only 9% of the papers published, over the 2001–2005 period it amounted to a significant 38.3%, placing it second in the list of the 26 lines analyzed and behind only the field of performance. In turn, although the analysis of the environment in general and its influence on strategy falls from 33% to 22.7% in the periods considered, the industry analysis remains stable with around 11.4% of the papers published on strategy. Nonetheless, the sum of the two constitutes a significant 34.1% of the papers; in other words, it continues to be a major topic.

The above data show that although the importance of internal factors and the RBV has grown in presence and magnitude, the focus on external factors, though declining slightly, has maintained a significant position in the research into strategic management. Many researchers of strategy have for some time wondered whether the resource-based view or paradigm has been exhausted, and whether the pendulum will once again swing to the outside of firms or whether it will remain much longer on the internal side. In our opinion, and based on the aforementioned and other similar data, research will not turn its attention away from either internal or external factors. What may be happening is that research in strategy is exploring new topics and approaches both in the external sphere and within firms themselves.

Proof of this is to be found in the fact that although the more traditional approach of industrial organization has been fading, recent years have seen the growing significance of approaches that stress the importance of external considerations. Such is the case of the institutional approach, whose analysis of the factors of success considers certain explanatory variables related to a firm›s institutional environment (Peng, 2002; Peng et al., 2009). This approach extends the industry›s traditional analysis by introducing new environmental variables, such as the laws, traditions or culture of a region or country, in order to understand the relationship between the environment and strategy. Furthermore, certain recent trends in the RBV seek to extend this approach to include environmental elements – "the demand side" (Adner and Zemsky, 2006; Priem et al., 2012).

On the external side of Fig. 1, a different perspective has been gathering steam that is rooted more in Austrian economics-based notions of entrepreneurship (Kirzner, 1997; Jacobson, 1992). This latter perspective pictures markets as a process in constant transformation due to the behavior of individual firms. Here, restrictions on firm behavior are not so much due to objective external constraints as to the absence of entrepreneurial knowledge, which can be overcome by direct market participation. As a result, firms must engage in entrepreneurial action aimed at the discovery of profit opportunities which are unknown a priori. Firm agency and competitive agility therefore become much more central to this perspective, with the focal lens of competitiveness shifting from advantage to opportunities. Together, these developments underlie the notion of strategic entrepreneurship (Ireland et al., 2003), which combines entrepreneurship – opportunity-seeking behavior – with strategic management, i.e., advantage-seeking behavior. Hence the reason we call this approach the "Austrian entrepreneurship-based view".

Within the internal side, other approaches have appeared alongside the RBV that also highlight internal aspects of the firm, such as dynamic capabilities (Teece et al., 1997), the knowledge-based view (Nonaka, 1994; Grant, 1996) or the more recent resource orchestration approach (Sirmon et al., 2011). Likewise, the approaches that focus on individual or group behavior – behavioral strategy – maintain a close relationship with the RBV, but are now looking for the factors of success at a different level of analysis within the firm (Powell et al., 2011).

THE TENSION BETWEEN MACRO AND MICRO LEVELS

Although it has been less developed in the literature, we can also identify a kind of pendulum between a macro or firm level of analysis and a micro level of analysis more closely related to group and/or individual behavior within the firm. Although it is generally accepted that the object of analysis in strategic management is the firm as a whole, be it multi-business, multi-market or multinational (Rumelt et al., 1994), the factors of success may be rooted at both the micro and macro levels. Yet it would be accurate to say that the latter have prevailed over the former in research to date.

In a similar vein to the internal–external pendulum, in the early 1960s and 1970s we come across major contributions within both the macro and micro domains. The founders of the strategy field analyzed the success of a firm as a whole, and approached the firm itself as the principal unit of analysis. Such is the case, for example, in the work of Chandler (1962) or Ansoff (1965). Others, rooted more in organizational theory, who made significant contributions at a more macro level include the work of Burns and Stalker (1961) and Lawrence and Lorsch (1967), among others. At the same time, major contributions from the sphere of organizational theory and behavior, such as the work of Cyert and March (1963) and March and Simon (1958), clearly at the micro level, already stand out in this first period and emerged as highly influential at the dawn of the discipline. Equally relevant is the attention paid to firm management through the seminal contributions made by, for example, Barnard (1938) or Selznick (1957), who exerted a considerable influence in these early years.

We may therefore contend that approaches or frameworks that analyze both the macro and the micro levels coexisted at this initial stage. From the 1980s onwards, the huge importance garnered by research in industrial economics shifted the focus of attention on to macro aspects, particularly the structure of industry. This dominance of external and macro aspects led to a steady decline in the influence of both micro and macro organizational approaches, as reported by Ramos-Rodríguez and Ruiz-Navarro (2004) in their study on the discipline's evolution.

As in the case of the internal–external pendulum, the 1980s witnessed the appearance of the economics of organizations such as agency theory and transaction cost economics. These perspectives were located more toward the midway point on the macro-micro pendulum, as they opened the corporate black box to analyze the contractual relationship between actors in the case of agency theory, or the relationships between firms and markets in the case of transaction cost economics.

The early 1990s saw the powerful emergence of the RBV that swung the momentum of the pendulum back toward a more macro approach in strategic management. The possession of strategically valuable resources and capabilities is the fundamental source of a firm›s competitive advantage. Yet although this approach continues to be present, it is now considered insufficient for advantage since it is not enough to know that strategic resources permit creating and sustaining a competitive advantage; the more important issue is to understand how.

Accordingly, the mid 1990s onwards see the appearance of approaches closely linked to the RBV, albeit at a lower level of analysis. Such is the case of the knowledge-based view that emphasizes the creation of intangible knowledge in a firm based on people and the relationships between them (Nonaka, 1994; Grant, 1996), or that of dynamic capabilities that deals with how to renew and improve resources on an ongoing basis to enable a firm to adapt to its environment (Teece et al., 1997; Wang and Ahmed, 2007). Along a different path, yet heading in the same direction, the approach that we might refer to as "resource orchestration" addresses the issue of how to create valuable capabilities and resources through the actions undertaken by a firm›s own management, thereby seeking to analyze the black box of the simple possession of resources and highlight the importance of their creation and management (Sirmon et al., 2007, 2011).

More recently, the line of research on microfoundations and behavioral strategy places emphasis on the individual level in order to analyze the behaviors that have an impact on strategy (Felin and Foss, 2005; Powell et al., 2011). The aim is to apply the knowledge of behavioral sciences, basically psychology and organizational behavior, to the issues of concern to strategic management. The overall purpose is to address and analyze strategy through realistic assumptions of human cognition, emotions and social behavior (Powell et al., 2011). These considerations connect with the RBV to the extent that they seek to analyze, among other aspects, why firms differ from one another (assumption of heterogeneity) or how organizational competencies and capabilities can be built up or generated on the basis of individual human resources. This approach, therefore, lays stress on a firm›s internal aspects while focusing attention on the influence of the individual level in firm success. Furthermore, it connects with the organizational approaches of the 1960s put forward by forerunners such as Cyert, March and Simon, whose contributions, as we have already stated, were highly influential in the discipline›s infancy.

On the right-hand side of the matrix in Fig. 1, we can see the shift in movement of the pendulum from the macro level of industrial organization toward a more micro level. Through the contributions made by the Austrian economists such as Kirzner (1973), which Jacobson (1992) terms the Austrian school of strategy, the emphasis shifts from the more macro level of the market or industry toward the more micro level of opportunities and firms' actions. Whereas in the industrial organization and institutional approaches the context is mostly taken as a given, in the Austrian economics approach the opportunities available are determined by the firms themselves through their actions and discoveries, irrespective of the industry and institutional context. Since market evolution is difficult to predict in advance, business opportunities are transient and need to be uncovered or generated by economic agents. As a result, the alert firm proactively engages in initiatives aimed at the discovery and the development of opportunities where existing competition is relatively limited or the value proposition is innovative and unique.

As the Austrian school of strategy underscores the importance of an entrepreneurial and action orientation within the firm, this approach begins to share a more micro-level perspective of strategy and firm behavior with micro-foundations and behavioral strategy. To elaborate further, one would expect entrepreneurial and action orientation to be reflected in a firm›s business model, or 'way of doing things', enabled by systems of activities and governance that emphasize quick decision-making, coordination and fast mobilization of efforts and resources (Doz and Kosonen, 2010). The role of the top management team in creating such an orientation would obviously be quite central here. To some extent then, even though the starting point of the Austrian school and microfoundations literature may differ, with the latter based more on cognition and behavioral strategy and the former more on Austrian economics, the two end up both complementing and reinforcing one another at the micro level.

To sum up, the two pendulums operate simultaneously and are in continuous movement, and at times even overlap one another. Consequently, we now encounter fully valid approaches that focus their attention both at macro level and at micro level, and within both the internal and external domains. To put it another way, research into strategic management has evolved toward a greater complexity of the issues addressed and approaches adopted. This means that one might expect a greater capacity for analysis of the complex issues it seeks to tackle, without focusing exclusively on one type of problem or one type of approach.

ABOUT THE MONOGRAPH

Bearing in mind the different contributions included in the special issue, its own title and that of this paper refers to the frontiers of research in strategic management. Indeed, it may be posited that, in one way or another, the works included here seek to explore that frontier.

First, drawing upon the reflections made in the previous section, we encounter three works that review the literature and analyze the issues from three distinct perspectives: external-macro (institutional approach), external-micro (action-based perspective) and internal-micro (microfoundations approach). Second, the frontier of strategy research is explored by addressing topics of special significance and growing interest in recent years. Such is the case with research into (a) multinational enterprises (MNEs) and the global context in which an ever-growing number of firms now operate; and (b) the relationship between a firm and its environment with respect to consideration of its social responsibility and the impact the firm has on the natural environment. Third, as noted earlier, recent years have seen a successful exploration and application of bibliometric techniques for analyzing the discipline of strategic management. Extending this from a methodological point of view, it is important to provide alternatives for the measurement of complex variables that allow progress to be made in empirical research.

Two of the articles address this important frontier issue in the MNE and the institutional contexts.

One of the foremost aspects of strategy research, partially responsible for the emergence and development of different theoretical approaches over the last few decades, has been the search for sources and reasons underlying firms' competitive advantage. Broadly speaking, extant frameworks that revolve around the sustainability of competitive advantage attribute a firm's superior performance to its positioning in the industry's structure and/or the possession of critical resources. Yet the logic implicit in these perspectives may well be out of tune with the current environment, characterized by more dynamic and complex behavior manifested by markets and firms, where competitive advantages are rendered obsolete faster than ever. The paper by Madhok and Marques (2014) proffers an alternative theoretical perspective to explain a firm's competitive behavior based on action. This approach revolves around a logic of action and stresses an entrepreneurial approach and firms' agility as the basis of their competitiveness. Such an action-based logic not only shifts the focus of attention away from industry position or the possession of resources as the singular explanation for a firm's competitiveness, but also provides new possibilities for analyzing how firms with fewer advantages can compete with already established firms.

Recent years have seen the emergence of a line of research in strategic management that focuses on an analysis of its microfoundations. This line highlights the need to consider aspects related to the actions of individuals and their interactions as independent variables underlying firm performance. The paper by Molina-Azorín (2014) conducts a systematic review of the existing literature on microfoundations, indicating their usefulness and main characteristics. It contributes to the literature by identifying the main areas of study, the benefits, opportunities and potential of the research stream in advancing the field of strategy, as well as certain limitations and challenges that need to be overcome. Furthermore, it analyzes the way in which micro and macro aspects may be integrated within strategy research, an issue of great interest and relevance for the progress of research in management and the study of organizations. To do so, it considers possible methodologies which may favor that integration and examines works that use a multilevel approach. These issues are analyzed mainly within the framework of the RBV, as the study of microfoundations seeks explanations for the heterogeneity of firms based on the differences between individuals.

Third, the paper by Garrido et al. (2014) analyzes the role played by institutions in answering questions of interest for strategic management from an institutional perspective. One of the main challenges that this perspective faces is to develop metrics that capture the institutional dimension more accurately. Their paper makes a major contribution in this regard by providing a detailed analysis of the main metrics used in strategic management research from the institutional perspective. The authors first review the main areas where institutional theory has been most widely developed and applied, such as diversification, foreign direct investment or modes of entry into foreign markets, and corporate governance, and then zero in on how this has been done, especially as regards the incorporation of the role of institutions within the empirical framework of the analyses. Their article therefore provides a guide that may serve as a reference for researchers when incorporating the institutional dimension into their empirical studies.

Fourth, one of the prominent emerging topics of strategic management research in recent years is corporate social responsibility. The paper by Gallardo-Vázquez and Sánchez-Hernández (2014) analyzes the extent to which the information on social responsibility in the hands of the managers of small and medium-size enterprises (SMEs) informs a positive predisposition toward the performance of responsible actions toward the natural environment. In particular, they verify whether there is a relationship between the components of the environmental responsibility of these firms, depending on whether the actions are more or less closely linked to their reporting. Curiously enough, SMEs are an important group for examining whether the more responsible firms are the ones that do the most reporting. This is because the responsible behavior of such firms tends to be driven more by the will of their managers rather than by image, and they tend to focus on internal stakeholders because they find it difficult to manifest their responsibility externally, especially in terms of environmental issues. Thus, within the empirical context of a Spanish region, Extremadura, the results of this study confirm that the greater the concern for gathering and receiving information on social responsibility, the greater the voluntary disposition toward responsible environmental management and the greater the importance attributed to actually disseminating the corporate responsibility practiced. Consequently, the efforts made by firm management, public authorities and business organizations in information, awareness or training in social responsibility have a positive effect on the attitude they manifest toward the environment. As can be seen, this approach suggests that the institutional context has a major influence on firms' strategic initiatives.

The fifth and last paper, by Dabic et al. (2014), reviews the research on MNEs and the application of strategic management approaches. For a long period, the research on the strategy of MNEs was rooted in internationalization theory. Based on transaction cost logic, it has difficulty in explaining how MNEs can build their competitive advantage. Developments in agency theory, the RBV, the knowledge-based approach and game theory have sought to complement and extend the explanations for the decisions, behavior and performance of these firms. As a result, there has been considerable dispersion and fragmentation in this line of research. The authors delve further into an understanding of how strategic thinking has impacted upon research into the strategy of MNEs. Following a review of the content of over a thousand papers published between 1975 and 2012, their paper maps the intellectual structure and its changes, concluding that the link between human capital and knowledge is essential for explaining how MNEs develop their strategies. The authors posit that this provides support for co-evolutionary theory, which involves the simultaneous development of arguments from different theories according to the research›s objectives and context. This theory is a promising line of research under the approaches based on resources and knowledge. Strategy›s contextual dependence suggests that different contexts require diverse approaches. This paper therefore concludes by suggesting that future lines of research should combine arguments from the various schools of strategic thinking.

CONCLUSION AND FINAL REFLECTIONS

Strategic management is a relatively young discipline that has been evolving on the back of numerous contributions from different fields in social sciences. Developments over the past few decades testify to the discipline's evolution and maturity, as it now enjoys a great capacity for analysis, as well as a greater array of theories, topics analyzed, and methodologies to do so. The metaphor of the pendulum that has been used to observe how the focus of attention has swung from the internal to the external domains and then back again to the internal one may be extended in two ways to better capture the discipline's evolution. First, besides the internal–external pendulum, a second pendulum is operational that swings between a macro and micro level approach toward key issues in strategic management. Second, these two pendulums are not currently at either end of the swing but are instead moving constantly and simultaneously.

Both the dimensions are present in past research on strategic management, and both are relevant today. This joint movement of the dual pendulum reflects the great complexity of strategic management as an academic discipline. In extending its capacity to understand and analyze the issues that it seeks to address, as well as to provide theories and tools for a deeper understanding of the issues, strategic management research needs to focus at the intersection of both the pendulums and the tensions this creates, since this is where the research frontier lies.

REFERENCES

1. Adner, R., Zemsky, P., 2006. A demand side perspective on sustainable competitive advantage. Strategic Management Journal 27 (3), 215---239.
2. Andrews, K., 1971. The Concept of Corporate Strategy. R.D. Irwin, Homewood.
3. Ansoff, H.I., 1965. Corporate Strategy: An Analytic Approach to Business Policy for Growth and Expansion. McGraw Hill, New York.
4. Azar, O.H., Brock, D.M., 2008. A citation-based ranking of Strategic Management journals. Journal of Economics and Management Strategy 17 (3), 781---802.
5. Barnard, C.I., 1938. The Functions of the Executive. Harvard University Press, Cambridge.
6. Barney, J.B., 1991. Firm resources and sustained competitive advantage. Journal of Management 17 (1), 99---120.

7. Burns, T., Stalker, G.M., 1961. The Management of Innovation. Oxford University Press, Oxford.
8. Chandler, A., 1962. Strategy and Structure: Chapters in the History of American Industrial Enterprise. MIT Press, Cambridge.
9. Cyert, R.M., March, J.G., 1963. A Behavioral Theory of the Firm. Prentice-Hall, Englewood Cliffs.
10. Dabic, M., Gonzalez-Loureiro, M., Furrer, O., 2014. Research on the strategy of multinational enterprises: key approaches and new avenues. Business Research Quarterly 17 (2), http://dx.doi.org/10.1016/j.brq.2013.09.001.
11. Doz, Y.L., Kosonen, M., 2010. Embedding strategic agility. Long Range Planning 43 (2---3), 370---382.
12. Felin, T., Foss, N., 2005. Strategic organization: a field in search of micro-foundations. Strategic Organization 3, 441---455.
13. Fredrickson, J.W. (Ed.), 1990. Perspectives on Strategic Management. Harper Business, New York.
14. Furrer, O., Thomas, H.A., Goussevskaia, A., 2008. The structure and evolution of the strategic management field: a content analysis of 26 years of strategic management research. International Journal of Management Reviews 10 (1), 1---23.
15. Gallardo-Vázquez, D., Sánchez-Hernández, M.I., 2014. Structural analysis of the strategic orientation to environmental protection in SME's. Business Research Quarterly 17 (2), http://dx.doi.org/10.1016/j.brq.2013.12.001.
16. García-Merino, T., Santos-Álvarez, V., 2009. Characterizing the top journals in strategic management: orientation, style, originality, and readability. Journal of Scholarly Publishing 40 (4), 420---439.
17. Garrido, E., Gómez, J., Maícas, J.P., Orcos, R., 2014. The institution-based view of strategy: how to measure it? Business Research Quarterly 17 (2), http://dx.doi.org/10.1016/j.brq. 2013.11.001.
18. Grant, R.M., 1996. Towards a knowledge-based theory of the firm. Strategic Management Journal 17 (Winter special issue), 109---122.
19. Guerras-Martín, L.A., Ronda-Pupo, G.A., 2013. Strategic management journal. In: Augier, M., Teece, D.J. (Eds.), The Palgrave

Encyclopedia of Strategic Management. Palgrave Macmillan, Houndmills.

20. Hoskisson, R.E., Hitt, M.A., Wan, W.P., Yiu, D., 1999. Theory and research in strategic management: swings of a pendulum. Journal of Management 25 (3), 417---456.
21. Ireland, R.D., Hitt, M.A., Sirmon, D.G., 2003. A model of strategic entrepreneurship: the construct and its dimensions. Journal of Management 29 (6), 963---989.
22. Jacobson, R., 1992. The 'Austrian' school of strategy. Academy of Management Review 17 (4), 782---807.
23. Jensen, M.C., Meckling, W.H., 1976. Theory of the firm: managerial behaviour, agency costs and ownership structure. Journal of Financial Economics 3, 305---360.
24. Ketchen, D., Boyd, B., Bergh, D., 2008. Research methodology in strategic management. Past accomplishments and future challenges. Organizational Research Methods 11 (4), 643---658.
25. Kirzner, I.M., 1973. Competition and Entrepreneurship. University of Chicago Press, Chicago.
26. Kirzner, I.M., 1997. Entrepreneurial discovery and the competitive market process: an Austrian approach. Journal of Economic Literature 35 (1), 60---85.
27. Lawrence, P., Lorsch, J., 1967. Organization and Environment: Managing Differentiation and Integration. Irwin, Homewood.
28. Madhok, A., Marques, R., 2014. Towards an action-based perspective on firm competitiveness. Business Research Quarterly 17 (2) , http://dx.doi.org/10.1016/j.brq.2014.03.002.
29. March, J.G., Simon, H.A., 1958. Organizations. Wiley, New York.
30. Mintzberg, H., 1990. The design school: reconsidering the basic premises of strategic management. Strategic Management Journal 11, 171---195.
31. Molina-Azorín, J.F., 2012. Mixed methods research in strategic management: impact and applications. Organizational Research Methods 15 (1), 33---56.
32. Molina-Azorín, J.F., 2014. Microfoundations of strategic management: toward micro-macro research in the resource-based

theory. Business Research Quarterly 17 (2), http://dx.doi.org/10.1016/j.brq.2014.01.001.

33. Nag, R., Hambrick, D.C., Chen, M.J., 2007. What is strategic management, really? Inductive derivation of a consensus definition of the field. Strategic Management Journal 28, 935---955.
34. Nerur, S.P., Rasheed, A.A., Natarajan, V., 2008. The intellectual structure of the strategic management field: an author cocitation analysis. Strategic Management Journal 29 (3), 319---336.
35. Nonaka, I., 1994. A dynamic theory of organizational knowledge creation. Organization Science 5 (1), 14---37.
36. Peng, M.W., 2002. Towards an institution-based view of business strategy. Asia Pacific Journal of Management 19 (2,3), 251---267.
37. Peng, M.W., Sun, S., Pinkham, B., Chen, H., 2009. The institutionbased view as a third leg for a strategic tripod. Academy of Management Perspectives 23 (3), 63---81.
38. Pettigrew, A., Thomas, H., Wittington, R. (Eds.), 2002a. Handbook of Strategy and Management. Sage, London.
39. Pettigrew, A., Thomas, H., Wittington, R., 2002b. Strategic management: the strengths and limitations of a field. In: Pettigrew, A., Thomas, H., Whittington, R. (Eds.), Handbook of Strategy and Management. Sage, London, pp. 3---30.
40. Porter, M.E., 1980. Competitive Strategy. Free Press, New York.
41. Powell, T.C., Lovallo, D., Fox, C.G., 2011. Behavioral strategy. Strategic Management Journal 32, 1369---1382.
42. Priem, R.L., Li, S., Carr, J.C., 2012. Insights and new directions from demand-side approaches to technology innovation, entrepreneurship, and strategic management research. Journal of Management 38 (1), 346---374.
43. Ramos-Rodríguez, A.R., Ruiz-Navarro, J., 2004. Changes in the intellectual structure of strategic management research: a bibliometric study of the Strategic Management Journal 1980---2000. Strategic Management Journal 25, 981---1004.
44. Ronda-Pupo, G.A., Guerras-Martin, L.A., 2010. Dynamics of the scientific community network within the strategic management field

through the Strategic Management Journal 1980---2009: the role of cooperation. Scientometrics 85 (3), 821---848.

45. Ronda-Pupo, G.A., Guerras-Martin, L.A., 2012. Dynamics of the evolution of the strategy concept 1962---2008: a co-word analysis. Strategic Management Journal 33 (2), 162---188.
46. Rumelt, R.P., Schendel, D.E., Teece, D.J. (Eds.), 1994. Fundamental issues in strategy. A research agenda. Harvard Business School Press, Boston.
47. Selznick, P., 1957. Leadership in Administration: A Sociological Interpretation. Harper & Row, New York.
48. Sirmon, D.G., Hitt, M.A., Ireland, R.D., 2007. Managing firm resources in dynamic environments to create value: looking inside the black box. Academy of Management Review 32 (1), 273---292.
49. Sirmon, D.G., Hitt, M.A., Ireland, R.D., Gilbert, B.A., 2011. Resource orchestration to create competitive advantage: breadth, depth, and life cycle effects. Journal of Management 37 (5), 1390---1412.
50. Teece, D.J., Pisano, G., Schuen, A., 1997. Dynamic capabilities and strategic management. Strategic Management Journal 18 (7), 509---533.
51. Wang, C.L., Ahmed, P.K., 2007. Dynamic capabilities: a review and research agenda. International Journal of Management Reviews 9 (1), 31---51.
52. Wernerfelt, B., 1984. A resource-based view of the firm. Strategic Management Journal 5, 171---180.
53. Williamson, O.E., 1975. Markets and Hierarchies: Analysis and Antitrust Implications. Free Press, New York.
54. Williamson, O.E., 1985. The Economic Institutions of Capitalism: Firms, Markets, Relational Contracting. Free Press, New York.

CHAPTER 11

THE INSTITUTION-BASED VIEW OF STRATEGY: HOW TO MEASURE IT

Elisabet Garrido[a], Jaime Gomez [b], Juan P. Maicas [c], Raquel Orcos [d]

[a] Universidad de Zaragoza, Violante de Hungría, 23, Zaragoza, Spain

[b] Universidad de La Rioja, Cigüena˜ 60, Logrono, ˜ Spain

[c] Universidad de Zaragoza, Gran Via 2, Zaragoza, Spain

[d] CEU San Pablo, Julián Romea, 23, Madrid, Spain

ABSTRACT

The institution-based view of strategy has emerged as a leading perspective in Strategic Management. It incorporates the institutional dimension when offering relevant answers to the fundamental questions of strategy. One of the challenges of this perspective is to develop stronger measures of institutions (Peng et al., 2009). This paper seeks to contribute in this direction by offering a detailed analysis of the main measures of institutions that previous works in Strategic Management have used. Our aim is to offer a guide that will help researchers to decide how they should incorporate the institutional dimension into their empirical work.

Keywords: Institution-based view, Formal institutions, Informal institutions, Institutional measures

INTRODUCTION

The field of Strategic Management has been traditionally governed by two big paradigms. While the 1980s were characterized by the dominance of the industry-based view (Porter, 1980), the 1990s were clearly the decade of the resource-based view of the firm (Barney, 1991). In recent decades, the two perspectives have moved like *swings of a pendulum* in the search for the answers to the fundamental questions of strategy (Hoskisson et al., 1999). In recent years, a third leading perspective seems to have conditioned the movement of the *pendulum* between these two important frameworks and has emerged in Strategic Management: the institution-based view (Peng et al., 2009). This perspective, with roots in both the economic – institutional economics – (North, 1990; Williamson, 1985) and sociological – institutional theory – literature (DiMaggio and Powell, 1983; Scott, 1995), incorporates the role of institutions in understanding why firms differ in terms of competitive advantage.

On the one hand, its emergence responds to external pressures from the new institutionalism movement in the social sciences in recent decades (North, 1990; Scott, 1995; Peng et al., 2009). On the other, its appearance seems to obey internal forces within the Strategic Management domain. In particular, the lack of attention to certain macro-contextual aspects by the industry and resource-based views has been important to promote the institutional perspective. The recognition of the relevance of institutions for competitive advantage is important so that they cease to be treated as background conditions or control variables (Peng et al., 2008). This recognition is also strongly related to Bamberger's (2008) claim for a more formal inclusion of contextual factors (e.g. institutions) in existing models that attempt to further advance management theories.

In the last decade, the institution-based view has been mainly used in the international business literature (Peng et al., 2008; Peng, 2009), particularly in the study of emerging economies (Cuervo-Cazurra and Genc, 2011; Cuervo-Cazurra, 2012). However, its use is increasing in Strategic Management in broader terms and some authors suggest that it should be understood as a complementary perspective to the industry- and resource-based views (Peng et al., 2009). This new paradigm is extremely young in the Strategic Management discipline and one of its challenges in the next few years is to demonstrate the importance of institutional factors for our understanding of competitive advantage.

It has been argued that, before its consolidation, the institution-based view needs to develop stronger measures of institutions (Peng et al., 2009). A first step in this direction would be to take stock of the available measures, analyzing their features and the interrelationships arising between them. Our objective in this paper is precisely this: to provide a guide that will help researchers to incorporate the institutional dimension of strategy into their empirical works by characterizing the different measures used in the literature. We first introduce the institution-based view of strategy and depict some of the areas in which the theory is more promising. Second, we focus on the practicalities of the application of this framework for the analysis of research questions in the Strategic Management arena. More precisely, we attempt to shed some light on the different alternatives available to scholars when incorporating the role of institutions into their empirical work. Although our intention is not to deal with *all* the measures that could be used to introduce the institution-based view of strategy, we analyze the wide variety of measurements that worldwide public sources offer to characterize the institutional environment. This analysis can be used as the benchmark against which new measures of institutions may be proposed in the future. We follow extant institutional studies and distinguish between formal and informal institutions (North, 1990), where the former are understood as laws, rules and regulations, and the latter as culture, norms and values. They are different but complementary sides of the coin that offer a complete picture of the institutional environment.

THE INSTITUTION-BASED VIEW OF THE FIRM

Definition and roots

Broadly speaking, institutions can be understood as the rules of the game in a society (North, 1990; Williamson, 1998). According to North (1991), they are the humanly devised constraints that structure political, economic and social interaction. They impose restrictions by defining legal, moral and cultural boundaries, setting off legitimate from illegitimate activities (Scott, 1995: 50). From a theoretical point of view, one of the major focuses of the literature on institutions has been on considering them as efficient solutions to problems of organization in a competitive framework (Williamson, 1975).

Although the early study of institutions can be traced back to the last decades of the 19th century, the emergence of the new institutionalism did not occur until the 1950s (Scott, 2008). As in Peng (2002), we use the term to refer to two independent streams of research that consider the role of institutions within society. The first, economic in nature, has its roots in the works of Coase (1937) and Williamson (1994, 1998). These authors considered formal and informal institutions as "background conditions" and focused on the determinants of choosing between different governance structures (e.g. markets vs. firms) to reduce transaction costs. By contrast, North (1990) mainly focused on formal and informal institutions as direct determinants of transaction costs in an economy. The evolutionary economics approach complements the static viewpoint of transaction cost economics by suggesting that firms continuously adapt their routines to the institutional framework through "an endogenous, experience-based learning process" (Knudsen, 1995: 203).[1]

Whereas the first stream of research has economic foundations and focuses on efficiency, the second viewpoint is sociological in nature, focuses on the role of legitimacy (Peng, 2002) and its main interest is to understand the role of institutions in explaining why organizations are similar. Under this perspective, the concept of isomorphism is based on the need to conform to what certain resource providers that are vital for organizational survival deem appropriate (Hambrick et al., 2005). Following DiMaggio and Powell (1983), pressure from societal institutions and collective rationality among organizations are the key elements to understand homogeneity. In other words, from this perspective, institutional support depends on a "general perception of organizational actions as desirable, proper or appropriate within some socially constructed system of norms, values, beliefs and definitions" (Suchman, 1995: 576).[2]

The institution-based view of strategy is the result of a consideration of both these streams of research in the context of business strategy. It conceives strategic choices as the result of the dynamic interaction between organizations and the formal and informal institutional environment (Peng, 2002).

The institution-based view of strategy

The field of Strategic Management has evolved a great deal since its inception. During the 1960s, the discipline paid attention to studies that focused on the importance of managerial skills (Andrews, 1971; Ansoff, 1965; Chandler, 1962). The 1980s were characterized by the role that industry structure played in determining differences in performance between firms, while the 1990s saw the birth of a new dominant paradigm, the resource-based view of the firm (Barney, 1991; Peteraf, 1993; Wernerfelt, 1984). The 21st century has *rescued* the environment as a key determinant in the analysis of firm behavior and performance. However, this analysis does not only focus on the industry-specific environment, as happened in the 1980s, but also highlights the institutional environment. As a consequence, recently a third leading perspective has emerged in strategic management. This perspective has been named the institution-based view (Peng et al., 2009).

One of the reasons that could explain the recent interest in this new paradigm is the globalization process that has affected most industries in the last years (Dunning and Lundan, 2008). This has also motivated an increase in academic research attempting to understand the origins of competitive advantage in international contexts. Although variations in the two components of the institutional environment, formal and informal, could be important within national contexts, it seems clear that its elements should present the highest differences when different countries are taken into account.

It is important to stress the intrinsic potential of this framework in order to answer relevant questions for business strategy. Previous literature has used some institutional dimensions before, but the link with what is called the institution-based view of strategy has only been implicit. Nevertheless, the use of the institution-based view of strategy, as an integrated framework for the understanding of competitive advantage, is more recent. In practical terms, this means that the variety of topics to which it has been applied and the depth of the analyses performed are still limited and call for further research. Following Bamberger (2008), the potential for contributions in this area should not just consist of the incorporation of control variables that help to provide an ad hoc understanding of some prevailing results in the literature, but should be an ex ante exercise in which the institutional framework is incorporated into the theories of strategic management.

The opportunity for this framework to interact with extant business strategy research can be assessed by considering the questions that have been deemed as important from the institution-based view (Peng et al., 2009), namely, (1) how do firms behave? (2) why are firms different? (3) what determines the scope of the firm?, and (4) what determines the success or failure of the firm in international competition? (Rumelt et al., 1994: 2). A first step in the search for the corresponding answers from the institution-based view has already started. Recent papers, since the seminal work of Peng (2002), have tried to analyze how institutions condition strategic choices (e.g. FDI location and entry mode, innovation adoption, executive compensation and entrepreneurship) as well as performance (e.g. from family and multinational firms). Table 1 shows several examples of the research questions that previous papers have tackled taking the institution-based view as a theoretical framework and some of their main findings.[3] Some of these questions are clearly related to the four broad areas delineated by Rumelt et al. (1994). For example, institutions have been related to the behavior of firms by considering how they determine entry mode choices, how they affect product diversification decisions or how they condition foreign direct investment locations. In the case of the latter, for instance, it has been found that, in contexts of low development of formal institutions, firms tend to prefer acquisitions and greenfield entries, rather than joint ventures (for more details, see Table 1).

Table 1: Examples of existing research questions and findings from the IBV.

Research questions	**References**	**Main findings**
How do institutions determine **entry mode choice** (greenfield vs. joint ventures vs. acquisitions)?	Estrin et al. (2009); Meyer et al. (2009); Santangelo and Meyer (2011)	In contexts of a lower development of formal institutions, the greenfield and acquisition are preferred to the joint venture (JV) entry mode. The preference for greenfield vs. cooperative entry modes increases with the distance between home and host countries in formal institutions. The preference for greenfield vs. cooperative entry modes has an inverted U-shaped relationship with the distance between home and host countries in informal institutions.
How do institutions affect **product diversification**?	Peng and Delios (2006); Wan (2005); Wan and Hoskisson (2003)	In contexts with weaker formal institutions, product diversification helps firms to internally create production factors whose acquisition in the market is subject to high transaction costs.
How do host market institutions determine **FDI location**?	Flores and Aguilera (2007); Quer et al. (2012); Kang and Jiang (2012)	Institutional factors better explain FDI location choice than economic factors. Multinationals prefer to enter in countries with lower institutional distance from their home countries. The presence of companies from the same home country overseas reduces the perceived institutional distance and favors multinational entry.
How do institutions influence **subsidiary performance** of MNEs?	Chan et al. (2008); Cuervo-Cazurra and Genc (2011); Ma et al. (2013); Chan et al. (2010)	Institutional development has a negative relationship with subsidiary performance. In contexts of lower institutional development, MNEs from developing countries have higher advantages than MNEs from developed countries. Subnational institutions also affect subsidiary performance.

How do institutions improve advantages from **innovation**?	Galang (2012); Zhu et al. (2012); Ahn and York (2011); Schneider et al. (2010)	Technological development and innovation advantages are favored by a higher efficiency of govern ment institutions and strong legal, educational and financial institutions, which reduce barriers to innovation.
How do institutions improve **entrepreneurship-** rates?	Lee et al. (2011); Stephan and Uhlander (2010); Peng et al. (2010)	Institutional contexts with a higher human orientation and assertiveness (informal institutions) and with entrepreneur-friendly bankruptcy laws (formal institutions) favor entrepreneurship.
How do institutions increase the effectiveness of **executive compensation** on performance?	Van Essen et al. (2012)	The positive effect of executive compensation on firm performance is higher when there are informal codes of good corporate governance as well as higher rule of law and investor protection.
How do institutions affect drivers of performance of **family business**?	Peng and Jiang (2010); Jiang and Peng (2011)	In contexts of a lower protection of minority shareholder rights, the presence of a family CEO positively influences firm performance, whereas the presence of a pyramidal structure has a negative influence.

Two main conclusions arise from the review of the literature that considers the role of institutions in the context of strategy. The first is that the theory suggests that the institution-based view of strategy could have a central role in the development of strategy research. For example, recent research equates the so-called influence rents with the four types of rents (monopolistic, efficiency, quasi-rents and Schumpeterian) on which research on strategic management articulates our comprehension of competitive advantage (Ahuja and Yayavaram, 2011). Influence rents are the "extra profits earned by an economic actor because the rules of the game of business are designed or changed to suit an economic actor or a group of economic actors" (Ahuja and Yayavaram, 2011: 1631). Therefore, they are the result of considering business strategy in the context of institutions from which economic rents may be extracted. This perspective on institutions seems to call for an even more central role of their interaction with business strategy, and it also suggests new relevant directions for research. Two examples of these questions are the following (Ahuja and Yayavaram, 2011). First, whether influence rents may provide firms implementing them with a sustainable improvement in performance, given the persistence of many institutions. Second, whether influence rents interact with the other four types of rents to provide the firm with competitive advantage.

Furthermore, our review also shows the variety of topics on which the institution-based view may be applied and the gap that exists between theory and empirics. Although it is not our objective to be exhaustive, some examples in which the theory could be applied to the analysis of topics of traditional interest for strategy researchers can be easily depicted. For example, the consideration of the institution-based view of strategy in the context of first-mover advantages could lead to an improved understanding of the advantages accruing to pioneers in an international context. This study could help to conciliate somc of the conflicting evidence on the topic by incorporating the macro-contextual factors determining performance to which we referred above. Another opportunity arises in multimarket competition theory. It is highly likely that the variation in the institutional characteristics of countries could affect the coordination capabilities of firms and, hence, the opportunities for mutual forbearance. For example, the different regulatory conditions prevailing in the different countries forming the market domains of firms or the cultural distances between these countries could affect the ability of multinationals to coordinate their international operations and their ability to collude with other multimarket firms (see, for example, Yu et al., 2009).

A second conclusion from our review is that the majority of papers focus on non-developed and emerging countries (China, India and Malaysia are three examples). This is unsatisfactory because, in this way, research on institutions renounces the exploration of a part of the variation of the underlying dimensions of institutions. The decomposition of the institutional environment, together with the combination of a larger set of different countries, could help in the task of effectively integrating the institution-based view into business strategy research. Ahuja and Yayavaram (2011: 1646) stress the possibility that (1) different sets of institutions may "be strong and well developed in a nation relative to other types of institutions in that nation" and (2) the level of development of a given set of institutions may vary across different nations. Therefore, opportunities arise for the exploitation of the sources of variation within and across different institutional contexts. Perhaps the most evident source of variation is the one that arises when considering the different countries in which firm strategies could be implemented. In this context, the comparison between developed, non-developed and emerging countries seems to be a fruitful avenue for research.

MEASURING INSTITUTIONS

Types of institutions

As mentioned above, our main objective in this paper is to provide a guide for measuring institutions by taking stock of the available measures. A first step in this direction is to distinguish between the different types of institutions that can be identified. We closely follow the widely accepted classification of North (1990), which distinguishes between formal and informal institutions.

Formal institutions refer to explicit rules in a society such as laws, regulations, property rights protection or discipline of economic and political markets and contracts (Dunning and Lundan, 2008; Meyer et al., 2009). They have to be explicitly established by an authority or an organization/individual. They are susceptible to change over time and can anticipate the desirable behavior of individuals and organizations from general and simple exchanges (e.g. through laws) to specific and complex ones (e.g. by a contract or a judicial resolution).

Informal institutions can be understood as those constraints that people in a society impose upon themselves to give a structure to their relations with others (North, 1990). These rules are transmitted from one generation to another by teaching and imitation (Boyd and Richerson, 1995) and, according to North (1990: 37), are a part of the heritage that we call culture. Tradition, religion, language, customs, values and trust-based relationships are some examples of informal institutions (Dunning and Lundan, 2008). Unlike formal institutions, informal institutions have their source in the values of a society and are difficult to change over time (North, 1990).

One of the main challenges for researchers attempting to integrate extant research on business strategy and the institution-based view of the firm is to obtain information on these two dimensions. In order to select the measures to analyze, we have imposed three criteria: (1) they have to be frequently used in the (Strategic) Management literature, (2) they also have to be publicly available for researchers and (3) they must have been specifically designed to capture the institutional characteristics of markets. In presenting the different alternatives, we offer a description of each index, its dimensions, time horizon and geographic scope, and we offer examples of papers that have incorporated these measures into their analyses. Additionally, wherever possible, correlation and factorial analyses are used for both formal and informal institutional indicators in order to determine the extent to which they converge or differ. We first analyze four measures that may be used to measure formal institutions and then move on to consider Hofstede's and the GLOBE studies as tools to capture informal institutions.

The measurement of formal institutions

In this section, we focus on four different approximations to the measurement of formal institutions: the *economic freedom index*, the *political constraints index*, the *corruption perceptions index* and the *worldwide governance indicators*. We first briefly present their main characteristics and then compare them.

The Index of Economic Freedom (EFI)

The Index of Economic Freedom (see Table 2 for more details), offered by the Heritage Foundation, measures the extent of government support to a person's control of the fruits of his/her labor and initiative. This index has been used in previous research as a measure of the *level of institutional development* of a country (Gubbi et al., 2010; Meyer and Sinani, 2009) or of the strength of market-supporting institutions (Meyer et al., 2009). A higher score in this index has been linked to stronger institutions, that is, those that support voluntary economic exchanges by assuring transparency and predictability.

Table 2: Index of Economic Freedom (EFI).

Organization	The Heritage Foundation
Source	http://www.heritage.org/index/
Researchers	Ambassador Terry Miller and Anthony B. Kim
Time availability	1995–2012 (with gaps)
Geographic scope	184 countries (2012)
Dimensions	10
Overall Index	Yes
Scale	0 (minimum freedom) to 100 (maximum freedom)

Countries with higher EFI: Hong Kong (89.9), Singapore (87.5) and Australia (83.1).

Countries with lower EFI: Cuba (28.3), Zimbabwe (26.3) and North Korea (1.00).

The overall index is calculated as the average of 10 dimensions that are grouped into four categories: (i) rule of law, (ii) regulatory efficiency, (iii) open markets and (iv) limited government. Rule of law refers to the extent to which the state protects private property rights from unfair expropriation and theft through the judicial system (*property rights*) and to which the market is free from political corruption through transparency in the regulatory procedures and processes (*freedom from corruption*). Regulatory efficiency determines whether the state allows individuals to establish and run their own businesses without interference (*business freedom*); whether it favors a policy based on low inflation, price stability and independence of the central bank (*monetary freedom*) and whether it allows individuals to work as much as they want and wherever they want and businesses to contract and dismiss workers according to activity requirements (*labor freedom*). The third category, open markets, describes the degree to which regulation supports market exchanges without restrictions in the national and international flow of goods and services (*trade freedom*) and capital (*investment freedom*). It also considers whether it encourages transparency, integrity and efficiency in the access to financing (*financial freedom*). Finally, limited government determines the level of lack of taxation (*fiscal freedom*) and spending (*government spending*) in a country. It is considered that high levels of government spending are associated with high levels of taxation, which reduces individuals' resources. For this reason, a high score in these two dimensions – lower levels of public taxation and spending – is considered a signal of greater freedom of individuals in a society, since they can freely decide the destination of their resources.

The Heritage Foundation offers an annual ranking of countries according to the overall EFI index. In addition, the Heritage Foundation classifies the listed countries according to five categories: free (100–80), mostly free (79.9–70), moderately free (69.9–60), mostly unfree (59.9–50) and repressed (49.9–0).

Table 3 provides the descriptive statistics for the overall index and for the ten dimensions it consists of (from 1995 to 2012). As it can be observed, the average score is higher for dimensions such as fiscal and monetary freedom (71 over 100), whereas the lowest average scores refer to property rights protection and freedom from corruption (48 and 40 over 100, respectively). Almost all the EFI dimensions present significant correlations within them. The correlations are especially high for the pairs (1) property rights and freedom from corruption (r = 0.83), (2) business freedom and property rights protection (r = 0.75) and (3) investment and financial freedom (r = 0.72). Interestingly, all the correlations are positive with the only exception of the ones involving fiscal freedom and government spending (limited government category), which present some negative correlations with the other EFI dimensions.

Table 3: Descriptive statistics EFI (1995–2012).

Index	Obs.	Mean	S.D.	Min	Max	1	2	3	4	5	6	7	8	9	10	11
1. Overall	2832	58.95	11.94	1	90.5	–										
2. Property rights	2832	48.72	24.02	0	95	0.79*	–									
3. Freedom from corruption	2832	40.34	23.51	4	100	0.75*	0.83*	–								
4. Fiscal freedom	2832	71.00	16.00	0	99.9	0.30*	−0.11*	−0.11*	–							
5. Government spending	2832	65.17	24.71	0	99.3	0.06*	−0.30*	−0.38*	0.47*	–						
6. Business freedom	2832	64.01	15.93	0	100	0.78*	0.75*	0.69*	0.05*	−0.17*	–					
7. Monetary freedom	2832	71.60	17.44	0	95.4	0.64*	0.39*	0.44*	0.20*	0.05*	0.38*	–				
8. Labor freedom	1342a	61.13	17.13	0	100	0.57*	0.39*	0.37*	0.20*	−0.03	0.50*	0.30*	–			
9. Trade freedom	2832	66.45	16.81	0	95	0.59*	0.37*	0.45*	0.21*	−0.15*	0.39*	0.34*	0.27*	–		
10. Investment freedom	2832	52.17	20.25	0	95	0.76*	0.67*	0.56*	−0.01	−0.13*	0.60*	0.38*	0.29*	0.39*	–	
11. Financial freedom	2832	50.39	20.22	0	90	0.80*	0.66*	0.60*	0.08*	−0.12*	0.61*	0.44*	0.34*	0.44*	0.72*	–

A Labor freedom was incorporated in 2005. Correlations are calculated for the period 2005–2012, with 1342 observations.

* $p < 0.01$.

Given the high correlations between some EFI dimensions, we have considered it interesting to carry out a factor and a cluster analysis with the aim of determining the extent to which the dimensions differ. Both techniques show similar results for all the years for which the information is available. Therefore, we base our comments on the most recent year (2012).[4] Using a principal component factor analysis with varimax rotation, two significant factors emerge (Table 4). All the dimensions have significant factor loadings (i.e. ≥± 0.40) on only one of the two factors. The first factor includes three categories, i.e. rule of law (property rights and freedom from corruption), regulatory efficiency (business, monetary and labor freedom) and open markets (trade, investment and financial freedom). The second factor includes the limited government category (fiscal freedom and government spending). After carrying out a cluster analysis using the ten dimensions, the countries are divided into two groups, in accordance with the Calinski-Harabasz test. When the average factor scores are calculated for the two groups, it is observed that countries in Group 1 are characterized by a higher score in Factor 1 (rule of law, regulatory efficiency and open markets categories) and a lower score in Factor 2 (limited government) than countries in Group 2.

Table 4: EFI categories, scores and country classification (2012).

Principal component factor analysis		
Variable	**Factor 1**	**Factor 2**
Property rights	**0.90**	−0.23
Freedom from corruption	**0.86**	−0.27
Fiscal freedom	−0.03	**0.86**
Government spending	−0.26	**0.70**
Business freedom	**0.82**	0.08
Monetary freedom	**0.65**	0.38
Labor freedom	**0.48**	0.37
Trade freedom	**0.71**	0.23
Investment freedom	**0.86**	0.00
Financial freedom	**0.86**	0.05

Cluster analysis		
Group	**Average factor score**	**Countries**
Group 1	***Higher av. score factor 1 = 1.10*** • Rule of law (property rights and freedom from corruption); • Regulatory efficiency (business, monetary and labor freedom); • Open markets (trade, investment and financial freedom). ***Lower av. score factor 2 = −0.32*** • Limited government (fiscal freedom and government spending)	Armenia, Australia, Austria, Bahrain, Barbados, Belgium, Botswana, Bulgaria, Canada, Chile, Colombia, Cyprus, Czech Republic, Denmark, Dominica, Estonia, Finland, France, Georgia, Germany, Greece, Hong Kong, Hungary, Iceland, Ireland, Israel, Italy, Jamaica, Japan, Jordan, Lithuania, Luxembourg, Macedonia, Malaysia, Malta, Mauritius, Mexico, Montenegro, New Zealand, Norway, Oman, Poland, Portugal, Santa Lucia, Saint Vincent and the Grenadines, Saudi Arabia, Singapore, Slovakia, Slovenia, Spain, Sweden, Switzerland, The Netherlands, Trinidad y Tobago, United Kingdom, United States.
Group 2	***Lower av. score factor 1 = −0.51*** • Rule of law (property rights and freedom from corruption); • Regulatory efficiency (business, monetary and labor freedom); • Open markets (trade, investment and financial freedom). ***Higher av. score factor 2 = 0.15*** • Limited government (fiscal freedom and government spending).	Albania, Algeria, Angola, Argentina, Azerbaijan, Bangladesh, Belarus, Belize, Benin, Bhutan, Bolivia, Bosnia and Herzegovina, Brazil, Burkina Faso, Burma, Burundi, Ivory Coast, Cambodia, Cameroon, Cape Verde, Central African Republic, Chad, China, Comoros, Costa Rica, Croatia, Cuba, DR Congo, Djibouti, Dominican Republic, Ecuador, Egypt, El Salvador, Equatorial Guinea, Eritrea, Ethiopia, Fiji, Gabon, Ghana, Guatemala, Guinea, Guinea-Bissau, Guyana, Haiti, Honduras, India, Indonesia, Iran, Kazakhstan, Kenya, Kiribati, Kuwait, Kyrgyz Republic, Laos, Latvia, Lebanon, Lesotho, Liberia, Libya, Macau, Madagascar, Malawi, Mauritania, Micronesia, Moldova, Mongolia, Morocco, Mozambique, Namibia, Nepal, Nicaragua, Niger, Nigeria, North Korea, Pakistan, Panama, Papua New Guinea, Paraguay, Peru, Qatar, R. Congo, Romania, Russia, Rwanda, Santo Tomé y Príncipe, Samoa, Senegal, Serbia, Seychelles, Sierra Leone, Solomon I., South Africa, Sri Lanka, Suriname, Swaziland, Syria, Taiwan, Tajikistan, Tanzania, Thailand, The Bahamas, The Gambia, The Philippines, Timor-Leste, Togo, Tonga, Tunisia, Turkey, Turkmenistan, Uganda, Ukraine, United Arab Emirates, Uzbekistan, Vanuatu, Venezuela, Vietnam

This index has frequently been used to measure formal institutions. Although there are examples in which the 10 dimensions have been employed at the same time, in other papers, only some of the dimensions of Factor 1 are introduced into the empirical analyses. For instance, Meyer and Sinani (2009) use the average score for the business freedom, trade freedom, investment freedom, financial freedom and property rights dimensions to analyze variations in spillovers from foreign direct investment (FDI) in developing and developed countries and transition economies. Similarly, Meyer et al. (2009) focus on emerging economies to analyze entry mode strategies followed by MNEs depending on the level of institutional development based on the five dimensions previously mentioned.

Political Constraints Index (POLCON)

The Political Constraints Index (see Table 5 for more details) measures the extent to which policy changes are constrained by institutional and political factors (Henisz, 2000). It is assumed that a higher number of constraints is positive, given that it promotes stability, reducing political hazards and, thus, uncertainty. For this reason, this index has been used to measure the variable *political stability* (Guler and Guillén, 2009; Vaaler and Schrage, 2009).

Table 5: Political Constraint Index (POLCON).

Source	http://www-management.wharton.upenn.edu/henisz/
Researcher	Witold Henisz
Time availability	1960–2007 (with gaps)
Geographic scope	192 countries (2007)
Dimensions	No
Overall Index	Yes (2)
Scale	0 (high level of political hazards) to 1 (low level of political hazards)

Countries with higher POLCON: Morocco (0.73) and Belgium (0.69).

Countries with lower POLCON: Cuba (0.00), Iran (0.00) and Somalia (0.00).

The POLCON database offers two measures of political constraints. The first, *POLCON*, identifies the degree of constraints on policy change using data on the number of independent veto points in the political system (executive, legislative, judicial and sub-federal branches of government) and the distribution of political preferences both across and within these branches (Henisz, 2000: 4). There is an alternative measure, *POLCONJ*, which also takes into account the alignment between the judicial and executive powers as well as the polarization or fractionalization of the courts. A high score in this index means that political hazards are low and, thus, firms are in a better position to anticipate market conditions.

As can be observed in Table 6, the average POLCON index has increased from 1960 to 2007, which means that, on average, the political hazard has decreased over time. There is also a high variability among countries: some of them score 0 (minimum), whereas the maximum score is 0.89. It can be highlighted that POLCONJ is very stable over time.

Table 6: Descriptive Statistics POLCON (1960–2007).[a]

Index	Period	Obs.	Mean	S.D.	Min	Max
POLCON	1960–1980	2816	0.22	0.31	0	0.89
POLCON	1981–2000	3195	0.33	0.34	0	0.89
POLCON	2001–2007	1167	0.42	0.32	0	0.89
POLCON	**1960–2007**	**7178**	**0.30**	**0.33**	**0**	**0.89**
POLCONJ	1960–1980	2495	0.12	0.22	0	0.86
POLCONJ	1981–2001	2327	0.13	0.22	0	0.85
POLCONJ	**1960–2001**	**4822**	**0.12**	**0.22**	**0**	**0.86**

Correlation between POLCON and POLCONJ is 0.79 ($p < 0.01$).

When management researchers use this indicator, they mainly focus on POLCON instead of POLCONJ, perhaps because the two are highly correlated (0.79) and POLCON has a notably higher number of observations. Previous studies have observed that the political stability of a country allows a higher rate of entry into a foreign market by American venture capital firms (Guler and Guillén, 2009). Political stability also moderates the relationship between the residual state ownership of a privatized firm and the financial performance associated with strategic decisions of privatizing firms, and between the time since privatization and this performance (Vaaler and Schrage, 2009).

Corruption Perceptions Index (CPI)

The Corruption Perceptions Index (see Table 7 for more details) measures the extent to which international businessmen and financial journalist perceive corruption in the public sector of a given country. It includes perceptions about the bribery of public officials, kickbacks in public procurement, and the effectiveness of public sector anti-corruption efforts. It considers corruption in a strict sense, without other dimensions such as political instability, and it has been understood as the most comprehensive quantitative indicator of cross-country *corruption* available (DiRienzo et al., 2007).

Table 7: Corruption Perceptions Index (CPI).

Organization	Transparency International
Source	http://www.transparency.org/research/cpi/overview
Researcher	Johann Graf Lambsdorff
Time availability	1995–2011 (with gaps)
Geographic scope	183 countries
Dimensions	No
Overall Index	Yes
Scale	0 (high level of corruption) to 10 (low level of corruption)

Countries with the highest level of corruption: North Korea (1.0) and Somalia (1.0).

Countries with the lowest level of corruption: New Zealand (9.5), Denmark (9.4) and Finland (9.4).

The Transparency International Society also offers a ranking of countries. However, it is important to consider the scores of each country instead of its position in the ranking in order to develop year-to-year comparisons (Lambsdorff, 2007) because this position not only depends on the scores of each country, but also on the number of participant countries in each edition. Since this number has varied over time, a country could occupy different positions in the ranking in two consecutive years although its score – and, thus, its perceived corruption – had not changed.

Table 8 shows the descriptive statistics for the CPI from 1995 to 2011. As can be observed, the average corruption index has decreased from the first period 1995–2000 to the second 2001–2011, meaning that, on average, the level of corruption in the public sector has increased. This decrease of the index is caused by the inclusion in this study of new countries with low scores (high level of corruption) in the CPI in the first decade of the 21st century. While the number of participant countries in 2000 was 91, it increased to 182 in 2011.

Table 8: Descriptive Statistics CPI (1995–2011).

Period	Obs.	Mean	S.D.	Min	Max
1995–2000	420	5.05	2.48	0.69	10
2001–2011	1665	4.13	2.19	0.40	9.9
1995–2011	**2085**	**4.32**	**2.28**	**0.4**	**10**

Unlike EFI and POLCON, the CPI is based on only one dimension/indicator. Previous studies have used it to analyze how the level of corruption of a country influences firm decisions. For instance, Habib and Zurawicki (2002) and Pajunen (2008) analyze developed and developing countries and find a negative impact of the CPI on the level of FDI in a country. Spencer and Gomez (2011), analysing Eastern European countries and Ghana, found a positive relationship between the host country corruption environment and the pressure that subsidiaries face to engage in bribery locally. Other studies have analyzed how the access to information and technology in a given country is negatively related to the level of corruption (DiRienzo et al., 2007), whereas the more rapid rate of change in FDI in a country is positively related to it (Robertson and Watson, 2004).

Worldwide Governance Indicators (WGI)

The Worldwide Governance Indicators (see Table 9 for more details) have been used in previous studies to measure the *governance* of countries, in other words, the set of traditions and institutions by which authority in a country is exercised and by which it is expected to affect economic activity (Cuervo-Cazurra and Genc, 2011; Globerman and Shapiro, 2003; Oh and Oetzel, 2011). It includes three categories of information: (1) the process by which governments are selected, monitored and replaced; (2) the capacity of the government to effectively formulate and implement sound policies; and (3) the respect of citizens and the State for the institutions that govern the economic and social interactions among them.

Table 9: Worldwide Governance Indicators (WGI).

Organization	World Bank Group
Source	http://info.worldbank.org/governance/wgi/index.asp
Researcher	Daniel Kaufmann, Aart Kraay, Massimo Mastrucci
Time availability	1996, 1998, 2000, 2002–2011
Geographic scope	215 countries
Dimensions	6
Overall Index	No
Scale	−2.5 (low governance) to 2.5 (high governance) (approx.)

There are six governance indicators that are based on the three previous categories: (1) *Voice and accountability* (VA) determine the degree to which citizens can exert their freedom rights and participate in selecting their government, while *Political stability and absence of violence/terrorism* (PS) measure perceptions about the likelihood that the government would be destabilized by unconstitutional or violent means or by terrorists; (2) *Government effectiveness* (GE) includes perceptions about the quality of public services, the civil service, policy formulation and the implementation and credibility of public policies, while *Regulatory quality* (RQ) captures the degree to which governments implement policies and regulations to promote private sector development; (3) *Rule of law* (RL) measures the quality of contract enforcement, property rights protection and the likelihood of crime and violence, and *Control of Corruption* (CC) measures the extent to which public power is exercised for private gain and the influence of private interests on public policies. Thus, in a country with high scores in these dimensions, economic agents can more easily anticipate the results of their economic activities, provided that there is a respect for regulation and contracts, a lack of violence and corruption, and regulatory quality and effectiveness.

Table 10 provides the descriptive statistics of WGI from 1996 to 2011. As can be observed, the average score is similar for all indicators (close to 0). All of them are highly correlated, with values ranging from 0.66 to 0.93.

Table 10: Descriptive Statistics WGI (1996–2011).

	Obs.	Mean	S.D.	Min	Max	1	2	3	4	5	6
1.VA	2683	−0.01	1.00	−2.30	1.83	–					
2.PS	2640	−0.03	1.00	−3.30	1.95	0.69*	–				
3.GE	2632	−0.01	1.00	−2.45	2.37	0.78*	0.69*	–			
4.RQ	2632	−0.01	1.00	−2.67	2.22	0.80*	0.66*	0.93*	–		
5.RL	2661	−0.02	1.00	−2.68	2.01	0.81*	0.79*	0.93*	0.89*	–	
6.CC	2635	−0.01	1.00	−2.06	2.59	0.77*	0.73*	0.93*	0.87*	0.93*	–

*$p < 0.01$.

To understand the structure of the indicators, we have performed an exploratory factor analysis using the principal component method with varimax rotation for the most recent period in which information is available (2011).[5] In this case, the six indicators were reduced to one factor, with factor loadings between 0.83 and 0.98.

Despite the high correlation between indicators, previous studies vary in the way that they have used them in empirical analyses. For instance, Globerman and Shapiro (2003) take the common factor based on the six indicators to show that governance institutions determine FDI inflows and outflows in developed and developing countries. Other researchers perform the empirical analysis for each of the six indicators, instead of reducing them to a common factor. For instance, Cuervo-Cazurra and Genc (2011) analyze developing country MNEs in comparison with developed country MNEs and find that the indicators related to regulatory quality, rule of law and control of corruption of the least developed countries have a key role in explaining the prevalence of developing-country MNEs among the largest subsidiaries of foreign firms in these countries. Oh and Oetzel (2011) also include each of the six indicators to analyze the moderating role of governance quality in the variation of European MNEs' investments at subsidiary level in response to major disasters. However, they find similar results for each of the six indicators.

Comparing the indicators

Having presented the main measures of formal institutions, in this subsection, we offer a comparison between them. Table 11 shows the correlations between all the previous indicators[6] for the 177 countries that appear in all the indexes for the last available year, 2007.[7]

Table 11: Correlation Matrix All Indicators (2007)**

		1	2	3	4	5	6	7	8	9	10	11	12	13	14	15	16	17	18	19	20	21	22
1	EFI	-																					
2	EFI_8_dimensions	0.96*	-																				
3	EFI_2_dimensions	-0.17	-0.43*	-																			
4	Property Rights	0.85*	0.91*	-0.47*	-																		
5	Freedom from Corruption	0.79*	0.87*	-0.54*	0.88*	-																	
6	Fiscal Freedom	-0.09	-0.30*	0.79*	-0.33*	-0.38*	-																
7	Government Spending	-0.18	-0.43*	0.93*	-0.46*	-0.53*	0.51*	-															
8	Business Freedom	0.83*	0.85*	-0.34*	0.75*	0.74*	-0.26*	-0.32*	-														
9	Monetary Freedom	0.58*	0.59*	-0.19	0.47*	0.44*	-0.18	-0.16	0.42*	-													
10	Labor Freedom	0.56*	0.52*	-0.02	0.39*	0.35*	0.03	-0.05	0.47*	0.18	-												
11	Trade Freedom	0.62*	0.65*	-0.29*	0.48*	0.51*	-0.13	-0.33*	0.50*	0.31*	0.20	-											
12	Investment Freedom	0.79*	0.83*	-0.37*	0.71*	0.63*	-0.29*	-0.34*	0.65*	0.51*	0.26*	0.53*	-										

13	Financial Freedom	0.77*	0.77*	-0.25*	0.60*	0.53*	-0.16	-0.25*	0.57*	0.48*	0.31*	0.52*	0.67*	-									
14	POLCON	0.57*	0.63*	-0.40*	0.58*	0.53*	-0.28*	-0.38*	0.56*	0.34*	0.23*	0.47*	0.63*	0.46*	-								
15	CPI	0.79*	0.89*	-0.58*	0.90*	0.98*	-0.42*	-0.57*	0.75*	0.48*	0.34*	0.57*	0.68*	0.55*	0.57*	-							
16	WGI	0.84*	0.92*	-0.54*	0.91*	0.92*	-0.36*	-0.55*	0.77*	0.52*	0.36*	0.58*	0.74*	0.61*	0.64*	0.94*	-						
17	Voice & Accountability	0.73*	0.80*	-0.50*	0.78*	0.70*	-0.39*	-0.47*	0.68*	0.45*	0.24*	0.53*	0.74*	0.66*	0.72*	0.74*	0.86*	-					
18	Political Stability	0.54*	0.61*	-0.43*	0.61*	0.68*	-0.21	-0.48*	0.45*	0.36*	0.18	0.43*	0.51*	0.38*	0.40*	0.68*	0.79*	0.62*	-				
19	Government Effectiveness	0.81*	0.89*	-0.51*	0.89*	0.92*	-0.34*	-0.52*	0.79*	0.49*	0.38*	0.55*	0.68*	0.54*	0.60*	0.93*	0.95*	0.75*	0.65*	-			
20	Regulatory Quality	0.90*	0.95*	-0.43*	0.88*	0.87*	-0.27*	-0.45*	0.79*	0.58*	0.40*	0.64*	0.80*	0.70*	0.65*	0.88*	0.94*	0.80*	0.63*	0.93*	-		
21	Rule of Law	0.81*	0.89*	-0.54*	0.91*	0.93*	-0.35*	-0.55*	0.76*	0.52*	0.37*	0.53*	0.69*	0.54*	0.60*	0.93*	0.97*	0.79*	0.76*	0.94*	0.90*	-	
22	Control of Corruption	0.78*	0.87*	-0.56*	0.90*	0.96*	-0.38*	-0.56*	0.73*	0.47*	0.35*	0.51*	0.66*	0.53*	0.57*	0.97*	0.95*	0.75*	0.70*	0.93*	0.86*	0.94*	-

**There are 177 common countries

*p<0.01; correlation > 0.70 in black.

As can be observed, there is a high correlation between EFI 8, CPI and WGI (0.89–0.94), which means that these indicators seem to measure related dimensions of formal institutions. Moreover, the correlation is especially high between some indicators and dimensions that capture similar institutional aspects. These include corruption (*freedom from corruption* from EFI, the CPI and *control of corruption* from WGI have a correlation that ranges from 0.96 to 0.98) and property rights protection (*property rights* from EFI and *rule of law* from WGI have a correlation of 0.91).

By contrast, correlations have a lower value between EFI 2 and the other indicators and dimensions, which seems to confirm that *fiscal freedom* and *government spending* capture government size instead of regulatory issues, as the other EFI dimensions do. The POLCON index also has a lower correlation with the other indicators, which suggests that it is measuring a different dimension.

Additionally, we have conducted an exploratory factor analysis through the principal component method for *EFI 8*, POLCON, CPI and the six WGI indicators in the last available period (2007). Table 12 shows that they are reduced to one factor that we have called *formal institutional development* (FID). We have calculated the value of this variable for each country. Table 12 offers a summary of the ten countries with the highest and lowest scores of the FID. As can be observed, the results are consistent with the classification obtained from the EFI dimensions and that presented in Table 4, because countries with the highest scores of FID belong to Group 1 whereas countries with the lowest score of FID belong to Group 2.

Table 12: Factor analysis (2007).

Variable	FID	Uniqueness
EFI 8	0.94	0.12
POLCON	0.71	0.50
CPI	0.96	0.08
VA	0.87	0.24
PS	0.74	0.45
GE	0.96	0.08

RQ	0.96	0.08
RL	0.97	0.05
CC	0.96	0.08

Highest FID		**Lowest FID**	
Country	**Factor 1**	**Country**	**Factor 1**
Denmark	2.19	Zimbabwe	−1.72
Finland	2.04	Turkmenistan	−1.60
New Zealand	2.01	Chad	−1.52
Switzerland	1.99	Guinea	−1.39
Sweden	1.99	Equatorial Guinea	−1.34
Iceland	1.94	Uzbekistan	−1.33
The Netherlands	1.92	Central African Rep.	−1.33
Australia	1.91	Côte d'Ivoire	−1.30
Austria	1.86	Venezuela	−1.29
Canada	1.85	Iran	−1.19

THE MEASUREMENT OF INFORMAL INSTITUTIONS

In this section, we review the most widespread alternatives to capture informal institutions: Hofstede and GLOBE (Global Leadership and Organizational Behavior Effectiveness). However, we have to acknowledge that there are also other options to measure informal institutions. For instance, Schwartz's value survey (1994), Inglehart and Associates' World Values Survey (1997) and Trompenaars and Hampden-Turner (1997) also allow the consideration of cultural issues across countries. In any case, due to their relevance and diffusion among the academic community (Hofstede, 2006; Tung and Verbeke, 2010; Smith, 2006), we focus our attention on Hofstede and GLOBE.

Hofstede

In his seminal book "*Culture's Consequences: International Differences in Work-Related Values*" (1980), Geert Hofstede developed several indicators to capture cross-national cultural differences. Hofstede initially collected data from 116,000 questionnaire-based surveys delivered to IBM managers participating in international management development courses in 1967 and 1973. At first, the study was conducted for 40 countries, but its coverage was extended to cover 93 countries in 2010 (Table 13).

Table 13: Hofstede's cross-cultural dimensions.

Source	http://geert-hofstede.com/national-culture.html
Researchers	Geert Hofstede, Michel Bond and Michel Minkov
Time availability	Scores are time invariant per dimension and country
Geographic scope	93 countries (2010)
Dimensions	6
Overall Index	No

The four initial dimensions were extended to five in 1991 and six in 2010 by Michel Bond and Michel Minkov, respectively. The six dimensions are the following: (1) *Power distance* measures the extent to which unequal distributions of power are accepted by the less powerful members of a society; (2) *Uncertainty avoidance*reflects societal stress as well as intolerance for uncertainty and ambiguity; (3) *Individualism* vs. *collectivism* shows the extent to which individuals are integrated into groups; (4) *Masculinity* vs. *femininity* is related to emotional gender roles in which masculinity is associated with assertiveness and competitiveness and femininity to modesty and caring; (5) *Long* vs. *short term orientation* is related to society's search for virtue, with short-term orientation being associated with respect for traditions, quick results and lower levels of savings and investments, and long-term orientation with a better adaptation to changes, long-term results and higher levels of savings and investments; (6) *Indulgence* vs. *restraint* determines the extent to which a society allows free gratification such as having fun (indulgence) or suppresses the gratification of needs by strict social norms (restraint). Therefore, a country with high scores in these six cultural dimensions would be characterized by (1) an unequal distribution of power; (2) intolerance of uncertainty and ambiguity; (3) individualism; (4) masculinity; (5) long-term orientation; and (6) indulgence.

If we calculate descriptive statistics for the six dimensions (Table 14), it can be observed that the average country has intermediate values. Interestingly there is only one significant correlation over 0.50 (the one between power distance and uncertainty avoidance, whose value is −0.61). This seems to imply that the dimensions are generally independent, in contrast to formal institutional indexes and dimensions, which were highly correlated in some cases.

Table 14: Descriptive statistics and correlations (Hofstede).[a]

	Obs.	Mean	S.D.	Min	Max	1	2	3	4	5	6
1.Power distance	71	59.11	21.65	11	104	–					
2.Uncertainty avoidance	71	44.39	24.00	6	91	−0.61*	–				
3.Individualism vs Collectivism	71	49.51	20.12	5	110	0.14	0.09	–			
4.Masculinity vs Femininity	71	67.55	23.48	8	112	0.23	−0.22	−0.07	–		
5.Long vs Short-term orientation	93	46.41	24.41	0	100	0.07	0.10	0.04	−0.02	–	
6.Indulgence vs. restraint	93	45.87	22.84	0	100	−0.27	0.19	0.11	−0.07	−0.39*	–

A We have exclusively focused on country scores and not on region scores.

*$p < 0.01$.

The use of these dimensions in empirical studies has focused on the role of culture in topics that belong to the Strategic Management domain and there have been two main approaches. The first focuses on analysing the impact of one or several dimensions on management behavior. For instance, Crossland and Hambrick (2012)study 15 developed countries and show that *individualism* in a country is positively related to the managerial discretion of the firms headquartered in that country, whereas *uncertainty avoidance* and *power distance* are negatively related to CEO discretion. Shao et al. (2013) show that the four initial dimensions of Hofstede moderate the relationship between employees' perceptions of justice and organizational outcomes such as satisfaction and commitment.

The second approach is based on analysing the effect of cultural indicators on firm strategies and outcomes by comparison. In this line the literature has used Hofstede›s dimensions to calculate different measures of *cultural distance*. One of the most frequent cultural distance measures based on Hofstede's dimensions is proposed by Kogut and Singh (1988),[8] and is calculated as follows:

$$CD_j = \frac{\sum_{i=1}^{6}((I_{ij} - I_{iu})^2)/V_i}{6}$$

where I_{ij} (I_{ij}) is the score of dimension i ($i = 1,\ldots,6$) for country $j(u)$ and V_i is the variance of dimension i. *Cultural distance* based on Hofstede›s indicators has been used in the analysis of a long list of research topics, mainly related to the internationalization process of multinationals. As examples, Chang and Rosenzweig (2001) show that joint ventures are preferred over greenfield and acquisition modes to enter into the United States when the cultural distance is high. Chakrabarti et al. (2009) analyze cross-border acquisitions involving 37 countries and show that cross-border acquisitions perform better in the long run when the acquirer and the target come from countries with higher cultural distance. As well as being a key independent variable, cultural distance between home and host countries has become one of the most frequent control variables in international studies that analyze MNEs' strategies and outcomes.

GLOBAL LEADERSHIP AND ORGANIZATIONAL BEHAVIOR EFFECTIVENESS (GLOBE)

The GLOBE project is a research program involving over 160 researchers from many parts of the world, conceived by Robert J. House in 1991 and initiated in 1993 (Table 15). It focuses on understanding the impact of culture on leadership and organizational processes. The project examines national cultures through a survey answered by middle managers from food processing, financial and telecommunications service industries and working in 62 countries (House et al., 2004, 2006).

Table 15: GLOBE dimensions.

Source	http://business.nmsu.edu/programs-centers/globe/instruments/
Researcher	Robert J. House
Time availability	Scores are time invariant per dimension and country
Geographic scope	62 countries
Dimensions	18 (9 of practices, 9 of values)
Overall index	No

GLOBE distinguishes cultural practices, that is, practices that individuals carry out *de facto* ("as is"), from cultural values, that is, practices that individuals should carry out ("should be"). There are nine dimensions each for both practices and values, which are described as follows: (1) *uncertainty avoidance* defines the extent to which societies avoid uncertainty by reliance on social norms; (2) *power distance* shows the degree to which members of a society agree that power should be unequally distributed; (3) *institutional collectivism* reflects the extent to which a society encourages collective actions, as well as the collective distribution of resources; (4) *in-group collectivism* considers the degree to which individuals express loyalty and cohesiveness in their families or organizations; (5) *gender egalitarianism* shows the degree to which a society minimizes gender discrimination; (6) *assertiveness* reflects the extent to which individuals are able to express their opinions; (7) *future orientation*considers the degree to which individuals are engaged in future-oriented behaviors; (8) *performance orientation* refers to the degree to which a society encourages individuals to improve their performance; and (9) *humane orientation* measures the extent to which a society or organization rewards individuals for being fair, altruistic and generous.

Table 16 shows the descriptive statistics and correlations of the cultural dimensions of practices and values considered by GLOBE. As can be observed, although GLOBE dimensions present higher significant correlations in comparison with Hofstede›s, they are generally moderate.

Table 16: Descriptive statistics and correlations (GLOBE).

Cultural practices														
	Obs.	**Mean**	**S.D.**	**Min**	**Max**	**1**	**2**	**3**	**4**	**5**	**6**	**7**	**8**	**9**
1.Uncertainty avoidance	62	4.17	0.60	2.88	5.37	1.00								
2.Future orientation	62	3.84	0.46	2.88	5.07	0.75*	1.00							
3.Power distance	62	5.14	0.45	3.59	5.80	−0.48*	−0.44*	1.00						
4.Institutional collectivism	62	4.24	0.43	3.25	5.22	0.38*	0.46*	−0.30	1.00					
5.Humane orientation	62	4.09	0.46	3.18	5.23	0.01	0.07	−0.15	0.41*	1.00				
6.Performance orientation	62	4.10	0.41	3.20	4.94	0.58*	0.62*	−0.33*	0.42*	0.25	1.00			
7.In-group collectivism	62	5.10	0.76	3.18	6.36	−0.59*	−0.40*	0.61*	−0.11	0.27	−0.11	1.00		
8.Gender egalitarianism	62	3.38	0.37	2.50	4.08	−0.06	−0.07	−0.32	−0.03	−0.14	−0.30	−0.23	1.00	
9.Assertiveness	62	4.13	0.37	3.38	4.89	−0.08	0.08	0.21	−0.38*	−0.42*	0.05	0.13	−0.10	1.00

Cultural values														
	Obs.	**Mean**	**S.D.**	**Min**	**Max**	**1**	**2**	**3**	**4**	**5**	**6**	**7**	**8**	**9**
1.Uncertainty avoidance	62	4.61	0.61	3.16	5.61	1.00								
2.Future orientation	62	5.44	0.51	2.95	6.20	0.64*	1.00							
3.Power distance	62	2.77	0.40	2.04	4.35	0.04	−0.36*	1.00						
4.Institutional collectivism	62	4.71	0.50	3.83	5.65	0.45*	0.50*	−0.37*	1.00					
5.Humane orientation	62	5.39	0.36	3.39	6.09	0.02	0.38*	−0.62*	0.07	1.00				
6.Performance orientation	62	5.88	0.56	2.35	6.58	0.25	0.69*	−0.61*	0.43*	0.61*	1.00			
7.In-group collectivism	62	5.64	0.41	4.06	6.52	0.35*	0.66*	−0.42*	0.36*	0.27	0.70*	1.00		
8.Gender egalitarianism	62	4.50	0.49	3.18	5.17	−0.49*	−0.16	−0.51*	0.00	0.28	0.28	0.21	1.00	
9.Assertiveness	62	3.83	0.64	2.66	5.56	0.19	0.03	0.29	−0.22	−0.12	−0.06	−0.02	−0.29	1.00

$*p < 0.01$.

GLOBE›s dimensions have been used by previous literature in similar ways to Hofstede›s indicators. For instance, Martin et al. (2007) analyze the impact of *performance orientation, human orientation* and *in-group collectivism* on local firm bribery activity, finding that only *in-group collectivism* is negatively related to it. Other studies calculate *cultural distance* through Kogut and Singh›s (1988) formula by using GLOBE dimensions and employ this variable for similar purposes as Hofstede›s indicators. As an example, Reus and Lamont (2009) focus on US multinational firms and show that, in international acquisitions, cultural distance based on GLOBE constrains communication between the acquirers and their acquired units but, at the same time, it provides more learning opportunities.

Hofstede vs. GLOBE

Since the publication of the GLOBE project, scholars have been heatedly debating the validity of Hofstede and GLOBE indexes, mainly from the perspective of the international business literature (Brewer and Venaik, 2011; Maseland and Van Hoorn, 2009; Smith, 2006). In spite of this discussion, there is no unanimity as to which index is superior. Tables 17a and 17b show correlations between Hofstede and GLOBE dimensions for both practices and values. The comparison of the two indexes shows that cultural dimensions that have similar names do not present high correlations and, apparently, are not measuring the same reality. As can be observed, the correlations are generally moderate and nonsignificant, although there are some exceptions. These exceptions sometimes coincide with dimensions that have similar meanings in Hofstede›s dimensions and GLOBE›s practices, such as (1) *uncertainty avoidance*(−0.68), (2) *individualism* vs. *collectivism* (−0.73) and (3) *power distance* (0.50), and in Hofstede and GLOBE›s values, such as (4) *future orientation* (−0.41) and (5) *individualism* vs. *collectivism* (−0.52). However, these high correlations are also found between pairs of dimensions that differ in their names. This is the case for (1) *uncertainty avoidance* (Hofstede) and *future orientation* (GLOBE practices) (−0.65) and for (2) *individualism* vs. *collectivism* (Hofstede) and *uncertainty avoidance*(GLOBE values) (−0.72), to give just a couple of examples. This situation has been highlighted by Javidan et al. (2006) as a response to the claim that GLOBE only replicates the five cultural dimensions presented by Hofstede (2006).

Table 17a: Correlations between cultural practices of GLOBE and Hofstede scores.[a]

		GLOBE								
		Cultural practices								
		1.Uncertainty avoidance	**2.Future orientation**	**3.Power distance**	**4.Institutional collectivism**	**5.Humane orientation**	**6.Performance orientation**	**7.In-group collectivism**	**8.Gender egalitarianism**	**9.Assertiveness**
Hofstede	1.Power distance	−0.41*	−0.26	0.50*	−0.16	0.10	−0.17	0.65*	0.04	−0.06
	2.Uncertainty avoidance	−0.68*	−0.65*	0.40*	−0.55*	−0.44*	−0.64*	0.22	0.00	0.16
	3.Individualism vs Collectivism	0.37	0.22	−0.47*	0.12	−0.14	0.10	−0.73*	0.18	0.04
	4.Masculinity vs Femininity	−0.13	−0.07	0.19	−0.11	−0.05	0.12	0.13	−0.15	0.35
	5.Long vs Short-term orientation	0.25	0.21	−0.21	0.29	−0.31	0.23	−0.19	0.01	−0.06
	6.Indulgence vs. restraint	0.18	0.24	−0.06	−0.03	−0.08	−0.11	−0.33	0.13	0.09

A Correlations referring to dimensions with similar names are highlighted to facilitate comparison. Correlations have been calculated for countries (and societies within a country) that are common for both indexes.

$*p < 0.01$.

Table 17b: Correlations between cultural values of GLOBE and Hofstede scores.[a]

		GLOBE								
		Cultural values								
		1.Uncertainty avoidance	**2.Future orientation**	**3.Power distance**	**4.Institutional collectivism**	**5.Humane orientation**	**6.Performance orientation**	**7.In-group collectivism**	**8.Gender egalitarianism**	**9.Assertiveness**
Hofstede	1.Power distance	0.66*	0.38*	0.01	0.33	−0.06	0.08	0.12	−0.42*	0.28
	2.Uncertainty avoidance	0.29	0.15	−0.31	0.32	−0.09	0.03	0.15	0.09	−0.39*
	3.Individualism vs Collectivism	−0.72*	−0.39*	0.14	−0.52*	0.09	−0.13	−0.23	0.35	−0.16
	4.Masculinity vs Femininity	0.08	0.11	0.09	0.01	0.01	0.03	−0.10	−0.05	0.17
	5.Long vs Short-term orientation	−0.26	−0.41*	0.16	−0.38*	−0.10	−0.46*	−0.59*	−0.09	0.08
	6.Indulgence vs. restraint	−0.21	0.03	−0.34	0.21	0.17	0.29	0.41*	0.54*	−0.31

A Correlations referring to dimensions with similar names are highlighted to facilitate comparison. Correlations have been calculated for countries (and societies within a country) that are common for both indexes.

- $p < 0.01$.

According to the evidence presented in this paper, it seems that both Hofstede and GLOBE differ in their measurement of the underlying variables and the literature has offered several reasons for these divergences. First, GLOBE considers a greater number of cultural dimensions than Hofstede, which, in practice, implies that one dimension of Hofstede might be explained by several dimensions of GLOBE. For instance, the *individualism* vs. *collectivism* of Hofstede would correspond to *institutional collectivism* and *in-group collectivism* in GLOBE (Venaik and Brewer, 2010). However, in the cases in which the name of the dimension and the underlying concept are the same, the way in which the question is posited is sometimes different. This is, for example, the case of the variable measuring *uncertainty avoidance* (Venaik and Brewer, 2010).

Second, Hofstede and GLOBE do not use the same scales and questions in their surveys. There are two main differences in this sense. Whereas GLOBE includes questions referring to both practices ("as is") and values ("should be"), Hofstede considers that practices are determined by values and, thus, only includes questions about values. Moreover, Hofstede questionnaires ask each individual about his/her personal perceptions about cultural dimensions, whereas GLOBE asks about society perceptions of cultural dimensions (Javidan et al., 2006; Maseland and Van Hoorn, 2009).

Third, Hofstede›s dimensions are based on surveys to managers of the same multinational (IBM), whereas GLOBE›s dimensions have been calculated on the basis of surveys to managers of more diverse origin. Moreover, the time of data collection differs between Hofstede and GLOBE. Whereas Hofstede started his project at the beginning of the 1970s and included new dimensions in 1991 and 2010, GLOBE started in the mid-1990s. Therefore, the cultural changes that might have taken place from the 1970s to the 1990s and from the 1990s to now could also explain the divergences between the two studies.

In sum, two main cultural indexes may be used by Strategic Management scholars to include informal institutions in their empirical analysis, namely, Hofstede›s and GLOBE›s. There is much discussion, based on their divergences, about which proposal is superior. These divergences can be explained by the structure of the survey, the different questions posited (even when the name of the variable is the same), the sample and the time at which these studies were conducted. Even admitting that these divergences exist, these indexes are frequently used in strategy journals. Some papers have even carried out analyses that take both measures into account in order to determine the robustness of their results (Hutzschenreuter and Voll, 2008; Hutzschenreuter et al., 2011).

CONCLUSIONS

The objective of this paper has been to provide a useful tool for strategy scholars interested in incorporating the institutional dimension of strategy into their research. Our main contribution has been to identify the different alternatives available to measure formal and informal institutions, the relationship between them and some guidance about their use in empirical work. We have focused on the information that is more easily available to the researcher, that provided by worldwide public sources. Although there is no doubt that this selection of these measures is, in some ways, arbitrary, it can be justified by their ability to capture the formal and informal dimensions into which the institutional characteristics of the environment may be divided, and by the frequency with which they have been used in the past.

The paper offers two main groups of conclusions. The first is related to the idea that the opportunities for research in the intersection between strategic management and the institution-based view seem to be promising. From the theoretical point of view, the institution-based view provides researchers with enough arguments for it to become a source of inspiration for Strategic Management, given that it directly tackles the questions that define its domain. Of special interest is its potential power to explain why firm performance differs. The idea that institutions not only condition the different solutions offered to organizational problems, but that they also may be influenced by firms to obtain "influence rents" also seems to contribute to that explanation, widening the applicability of the theory. Whether influence rents will become the "standard fifth type" of rents, at the same level as efficiency, monopolistic, Schumpeterian or quasi-rents, is a question that will remain open for some time and that also depends on the interest the subject raises. The fact that all firms are embedded in societal systems also seems to evidence its potential applicability.

From the empirical point of view, although with some exceptions, the existing evidence has centered on the analysis of international businesses. Although there is no doubt that the international context is one of the natural places in which the theory could be applied, the literature suggests other avenues for research as may also be derived from the analysis above. One of them is to further explore the variation in institutions between different countries, even if the firms considered are not executing an international business strategy. Up to now, research has focused on emerging economies, losing the opportunity to explore the differences between developed and non-developed countries. These differences not only affect our understanding of the strategies and performance of international firms when going abroad, but they also help us to understand why similar strategies may offer different results in different institutional contexts. In this line, we have observed that the institution-based view has already been used in a wide variety of contexts seeking to analyze, for instance, the role of institutions in explaining product diversification or entry mode strategies. However, there is still a range of topics to which the theory may be applied. This paper suggests the analysis of first-mover advantages or multimarket competition theory as future arenas where the intersection between existing theories and the institution-based view could give us a more nuanced picture of these phenomena.

The second group of conclusions relates to the main objective of the paper, which has been to provide researchers with a view of the variables that capture the institutional environment within which firms operate and that may condition their performance. First, we have provided clear evidence of the feasibility of measuring the different dimensions that define the institutional environment, given the existence of publicly available information. Second, we have studied a selected set of variables that refer to the two main types of institutions that are frequently distinguished, offering a brief description of each of them and analysing their interrelations. Regarding formal institutions, we have focused our attention on the Index of Economic Freedom, the Political Constraints Index, the Corruption Perception Index and the Worldwide Governance Indicators. The correlations between some of these four indexes and, sometimes, between their components are high enough to make us think that the underlying concept they are measuring is very similar.

Importantly, when a factorial analysis is performed, all the variables are reduced to one factor, apparently indicating that they are measuring the same underlying construct. From this evidence, our advice to researchers is to bear in mind that when measuring formal institutions, most of the analyzed alternatives point in the same direction.

A similar analysis has been performed to understand the variables proposed to measure the informal dimension of institutions. In this case, the sources of the data are Hofstede's study of culture and the GLOBE project. The main issue to highlight is that, although both studies attempt to measure the same underlying dimension, they achieve this task very differently and with a certain degree of contradiction. First, they differ in the number of underlying dimensions, the way the different questions are posited, the sources of the information provided and the timing. Second, and perhaps most important in terms of potential confusion, the dimensions that are similarly named are sometimes very difficult to conciliate and seem to be measuring different underlying dimensions. Our comparison of the Hofstede and GLOBE measures not only confirms this by sometimes even showing negative correlations between similarly named dimensions, but also that there are some unexpectedly high correlations between apparently different measures. Therefore, researchers incorporating the institution-based view are advised to be very careful when selecting the variable of informal institutions that best fits their theory. We suggest going directly to the source, either Hofstede or GLOBE, and understanding the meaning underlying each dimension in order to be more precise when selecting the measure. It is not only important to empirically justify the choice of one particular alternative, but also to theoretically argue in favor of the relationship between the theory used and the informal dimension analyzed.

In sum, this paper aimed to contribute to the consolidation of the institution-based view by presenting a guide for researchers for incorporating the institutional dimension into their works. This is just a first attempt; there is still a long way to go in integrating the extant Strategic Management theories and this new and promising perspective.

REFERENCES

1. Ahn, M., York, A., 2011. Resource-based and institution-based approaches to biotechnology industry development in Malaysia. Asia Pacific Journal of Management 28 (2), 257---275.
2. Ahuja, G., Yayavaram, S., 2011. Explaining influence rents: the case for an institutions-based view of strategy. Organization Science 22 (6), 1631---1652.
3. Andrews, K., 1971. The Concepts of Corporate Strategy. Dow JonesIrwin, Homewood, IL.
4. Ansoff, H., 1965. Corporate Strategy. McGraw Hill, New York.
5. Bamberger, P., 2008. Beyond contextualization: using context theories to narrow the micro-macro gap in management research. Academy of Management Journal 51 (5), 839---846.
6. Barney, J., 1991. Firm resources and sustained competitive advantage. Journal of Management 17 (1), 99---120.
7. Boyd, R., Richerson, P.J., 1995. Why does culture increase human adaptability? Ethology and Sociobiology 16 (2), 125---143.
8. Brewer, P., Venaik, S., 2011. Individualism-collectivism in Hofstede and GLOBE. Journal of International Business Studies 42 (3), 436---445.
9. Chakrabarti, R., Gupta-Mukherjee, S., Jayaraman, N., 2009. MarsVenus marriages: culture and cross-border M&A. Journal of International Business Studies 40 (2), 216---236.
10. Chan, C.M., Isobe, T., Makino, S., 2008. Which country matters? Institutional development and foreign affiliate performance. Strategic Management Journal 29 (11), 1178---1205.
11. Chan, C.M., Makino, S., Isobe, T., 2010. Does subnational region matter? Foreign affiliate performance in the United States and China. Strategic Management Journal 31 (11), 1226---1243.
12. Chandler, A.D., 1962. Strategy and Structure. MIT Press, Cambridge.
13. Chang, S.J., Rosenzweig, P.M., 2001. The choice of entry mode in sequential foreign direct investment. Strategic Management Journal 22 (8), 747---776.
14. Coase, R.H., 1937. The Nature of the Firm. Economica N.S.4, 385---405.

15. Crossland, C., Hambrick, D.C., 2012. Differences in managerial discretion across countries: how nation-level institutions affect the degree to which CEOs matter. Strategic Management Journal 32 (8), 797---819.
16. Cuervo-Cazurra, A., 2012. Extending theory by analyzing developing country multinational companies: solving the Goldilocks debate. Global Strategy Journal 2 (3), 153---167.
17. Cuervo-Cazurra, A., Genc, M., 2011. Transforming disadvantages into advantages: developing-country MNEs in the least developed countries. Journal of International Business Studies 39 (6), 957---979.
18. DiMaggio, P.J., Powell, W.W., 1983. The iron cage revisited: institutional isomorphism and collective rationality in organizational fields. American Sociological Review 48 (2), 147---160.
19. DiRienzo, C.E., Das, J., Cort, K.T., Burbridge, J., 2007. Corruption and the role of information. Journal of International Business Studies 38 (2), 320---332.
20. Dunning, J., Lundan, S., 2008. Institutions and the OLI paradigm of the multinational enterprise. Asia Pacific Journal of Management 25 (4), 573---593.
21. Estrin, S., Baghdasaryan, D., Meyer, K.E., 2009. The impact of institutional and human resource distance on international entry strategies. Journal of Management Studies 46 (7), 1171---1196.
22. Flores, R.G., Aguilera, R.V., 2007. Globalization and location choice: an analysis of US multinational firms in 1980 and 2000. Journal of International Business Studies 38, 1187---1210.
23. Galang, R.M.N., 2012. Government efficiency and International technology adoption: the spread of electronic ticketing among airlines. Journal of International Business Studies 43 (7), 631---654.
24. Globerman, S., Shapiro, D., 2003. Governance infrastructure and US foreign direct investment. Journal of International Business Studies 34 (1), 19---39.
25. Greenwood, R., Hinings, C.R., 1996. Understanding radical organizational change: bringing together the old and the new instituionalism. Academy of Management Review 21 (4),

26. 1022---1054.
27. Gubbi, S.R., Aulakh, P.S., Ray, S., Sarkar, M., Chittoor, R., 2010. Do international acquisitions by emerging-economy firms create shareholder value? The case of Indian firms. Journal of International Business Studies 41 (3), 397---418.
28. Guler, I., Guillén, M.F., 2009. Institutions and the internationalization of US venture capital firms. Journal of International Business Studies 41 (2), 185---205.
29. Habib, M., Zurawicki, L., 2002. Corruption and foreign direct investment. Journal of International Business Studies 33 (2), 291---307.
30. Hambrick, D.C., Finkelstein, S., Cho, T.S., Jackson, E.M., 2005. Isomorphism in reverse: institutional theory as an explanation for recent increases in intraindustry heterogeneity and managerial discretion. In: DiMaggio, P.J., Powell, W.W. (Eds.), Research in organizational behavior, vol. 26. JAI Press, Greenwich, CT, pp. 307---350.
31. Henisz, W.J., 2000. The institutional environment for economic growth. Economics & Politics 12 (1), 1---31.
32. Hofstede, G., 1980. Culture's Consequences: International Differences in Work-Related Values. Sage, Beverly Hills, CA.
33. Hofstede, G., 2006. What did GLOBE really measure? Researchers' minds versus respondents' minds. Journal of International Business Studies 37 (6), 882---896.
34. Hofstede, G., 2010. The GLOBE debate: back to relevance. Journal of International Business Studies 41 (8), 1339---1446.
35. Hoskisson, R.E., Hitt, M.A., Wan, W.P., Yiu, D., 1999. Theory and research in strategic management: swings of a pendulum. Journal of Management 25 (3), 417---456.
36. House, R.J., Hanges, P.J., Javidan, M., Dorfman, P.W., Gupta, V. (Eds.), 2004. Culture, Leadership and Organizations: The GLOBE Study of 62 Societies. SAGE Publications Inc., London.
37. House, R., Javidan, M., Dorfman, P., Sully de Luque, M., 2006. A failure of scholarship: response to George Graen's critique of GLOBE. Academy of Management Perspectives 20 (4), 102---114.

38. Hutzschenreuter, T., Voll, J.C., 2008. Performance effects of "added cultural distance" in the path of international expansion: the case of German multinational enterprises. Journal of International Business Studies 39, 53---70.
39. Hutzschenreuter, T., Voll, J., Verbeke, A., 2011. The impact of added cultural distance and cultural diversity on international expansion patterns: a Penrosean perspective. Journal of Management Studies 48 (2), 305---329.
40. Inglehart and Associates, 1997. World Values Survey. www.worldvaluesurvey.org
41. Javidan, M., House, R.J., Dorfman, P.W., Hanges, P.J., Sully de Luque, M., 2006. Conceptualizing and measuring cultures and their consequences: a comparative review of GLOBE and Hofstede approaches. Journal of International Business Studies 37, 897---914.
42. Jiang, Y., Peng, M.W., 2011. Are family ownership and control in large firms good, bad, or irrelevant? Asia Pacific Journal of Management 28 (1), 15---39.
43. Kang, Y., Jiang, F., 2012. FDI location choice of Chinese multinationsls in East and Southeast Asia: traditional economic factors and institutional perspective. Journal of World Business 47 (1), 45---53.
44. Knudsen, C., 1995. Theories of the firm, strategic management, and leadership. In: Montgomery, C.A. (Ed.), Resource-based and Evolutionary Theories of the Frim: Toward A Synthesis. Kluwer, Boston, pp. 179---217.
45. Kogut, B., Singh, H., 1988. The effects of national culture on the choice of entry mode. Journal of International Business Studies 19 (3), 411---432.
46. Lambsdorff, J.G., 2007. The Methodology of the Corruption Perceptions Index 2007. Transparency Internation., Berlin www. transparency. Org/content/download/23965/358196
47. Lee, S.-H., Yamakawa, Y., Peng, M.W., Barney, J.B., 2011. How do bankruptcy laws affects entrepreneurship development around the world? Journal of Business Venturing 26 (5), 505---520.
48. Ma, X., Tong, T.W., Fitza, M., 2013. How much does subnational regiona matter to foreign subsidiary performance? Evidence from

Fortune Global 500 Corporations' investments in China. Journal of International Business Studies 44 (1), 68---87.

49. Martin, K.D., Cullen, J.B., Johnson, J.L., Parboteeah, K.P., 2007. Deciding to bribe: a cross-level analysis of firm and home country influences on bribery activity. Academy of Management Journal 50 (6), 1401---1422.
50. Maseland, R., Van Hoorn, A., 2009. Explaining the negative correlation between values and practices: a note on the Hofstede-GLOBE debate. Journal of International Business Studies 40, 527---532.
51. Meyer, K.E., Estrin, S., Bhaumik, S.K., Peng, M.W., 2009. Institutions, resources, and entry strategies in emerging economies. Strategic Management Journal 30 (1), 61---80.
52. Meyer, K.E., Sinani, E., 2009. When and where does foreign direct investment generate positive spillovers? A meta-analysis. Journal of International Business Studies 40 (7), 1075---1094.
53. North, D.C., 1990. Institutions, Institutional Change, and Economic Performance. Harvard University Press, Cambridge, MA.
54. North, D.C., 1991. Institutions. The Journal of Economic Perspectives 5 (1), 97---112.
55. Quer, D., Claver, E., Rienda, L., 2012. Political risk, cultural distance, and outward foreign direct investment: empirical evidence from large Chinese firms. Asia Pacific Journal of Management 29 (4), 1089---1104.
56. Oh, C.H., Oetzel, J., 2011. Multinationals' response to major disasters: how does subsidiary investment vary in response to the type of disaster and the quality of country governance? Strategic Management Journal 32 (6), 658---681.
57. Pajunen, K., 2008. Institutions and inflows of foreign direct investment: a fuzzy-set analysis. Journal of International Business Studies 39 (4), 652---669.
58. Peng, M.W., 2002. Towards an institution-based view of business strategy. Asia Pacific Journal of Management 19 (2/3), 251---267.
59. Peng, M.W., 2009. Global Business 2009 Update. South-Western Cengage Learning, Mason USA.

60. Peng, M.W., Delios, A., 2006. What determines the scope of the firm over time and around the world? An Asia Pacific perspective. Asia Pacific Journal of Management 23 (4), 385---405.
61. Peng, M.W., Jiang, Y., 2010. Institutions behind family ownership and control in large firms. Journal of Management Studies 47 (2), 253---273.
62. Peng, M., Sun, S., Pinkham, B., Chen, H., 2009. The institutionbased view as a third leg for a strategy tripod. The Academy of Management Perspectives 23 (3), 63---81.
63. Peng, M.W., Wang, D.Y.L., Jiang, Y., 2008. An institution-based view of international business strategy: a focus on emerging economies. Journal of International Business Studies 39 (5), 920---936.
64. Peng, M.W., Yamakawa, Y., Lee, S.H., 2010. Bankruptcy laws and entrepreneur-friendliness. Entrepreneurship: Theory and Practice 34 (3), 517---530.
65. Peteraf, M.A., 1993. The cornerstones of competitive advantage: a resource-based. Strategic Management Journal 14 (3), 179---191. Porter, M.E., 1980. Competitive Strategy. The Free Press, New York.
66. Reus, T.H., Lamont, B.T., 2009. Double-edged sword of cultural distance in international acquisitions. Journal of International Business Studies 40 (8), 1298---1316.
67. Roberts, P.W., Greenwood, R., 1997. Integrating transaction cost and institutional theories: toward a constrained-efficiency framework for understanding organizational design adoption. Academy of Management Review 22 (2), 346---373.
68. Robertson, C.J., Watson, A., 2004. Corruption and change: the impact of foreign direct investment. Strategic Management Journal 25 (4), 385---396.
69. Rumelt, R., Schendel, D., Teece, D. (Eds.), 1994. Fundamental Issues in Strategy: A Research Agenda. Harvard Business School Press, Boston.
70. Santangelo, G.D., Meyer, K.E., 2011. Extending the internationalization process model: increases and decreases of

MNE commitment in emerging economies. Journal of International Business Studies 42 (7), 894---909.

71. Schneider, M.R., Schulze-Bentrop, C., Paunescu, M., 2010. Mapping the institutional capital of high-tech firms: a fuzzy-set analysis of capitalist variety and export performance. Journal of International Business Studies 41 (2), 246---266.
72. Schwartz, S.H., 1994. Beyond individualism---collectivism: new cultural dimensions of values. In: Kim, U., Triandis, H.C., Kagitcibasi, C., Choi, S.-C., Yoon, G. (Eds.), Individualism and collectivism: Theory, method and applications. , pp. 85---119.
73. Shao, R., Rupp, D.E., Skarlicki, D.P., Jones, K.S., 2013. Employee justice across countries: a meta-analytic review. Journal of Management 39 (1), 263---301.
74. Scott, W.R., 1995. Institutions and Organizations. Sage, Thousand Oaks, CA.
75. Scott, W.R., 2008. Institutions and Organizations: Ideas and Interest, 3rd ed. Sage, Thousand Oaks, CA.
76. Smith, P., 2006. When elephants fight, the grass gets trampled: the GLOBE and Hofstede projects. Journal of International Business Studies 37 (6), 915---921.
77. Spencer, J., Gomez, C., 2011. MNEs and corruption: the impact of national institutions and subsidiary strategy. Strategic Management Journal 32 (3), 280---300.
78. Stephan, U., Uhlander, L.M., 2010. Performance-based vs socially supportive culture: a cross-national study of descriptive norms and entrepreneurship. Journal of International Business Studies 41 (8), 1347---1364.
79. Suchman, M.C., 1995. Managing legitimacy: strategic and institutional approaches. Academy of Management Review 20 (3), 571---610.
80. Trompenaars and Hampden Turner, 1997. Ecological (cross-cultural) Database. www.thtconsulting.com
81. Tung, R.L., Verbeke, A., 2010. Beyond Hofstede and GLOBE: improving the quality of cross-cultural research. Journal of International Business Studies 41 (8), 1259---1274.

82. Vaaler, P.M., Schrage, B.N., 2009. Residual state ownership, policy stability and financial performance following strategic decisions by privatizing telecoms. Journal of International Business Studies 40 (4), 621---641.

83. Van Essen, M., Heugens, P.P., Otten, J., Van Oosterhout, J., 2012. An institution-based view of executive compensation: a multilevel meta-analytic test. Journal of International Business Studies 43 (4), 396---423.

84. Venaik, S., Brewer, P., 2010. Avoiding uncertainty in Hofstede and GLOBE. Journal of International Business Studies 41 (8), 1294---1315.

85. Wan, W.P., 2005. Country resource environments, firm capabilities, and corporate diversification strategies. Journal of Management Studies 42 (1), 161---182.

86. Wan, W.P., Hoskisson, R.E., 2003. Home country environments, corporate diversification strategies, and firm performance. Academy of Management Journal 46 (1), 27---45.

87. Wernerfelt, B., 1984. A resource-based view of the firm. Strategic Management Journal 5 (2), 171---180.

88. Williamson, O., 1975. Markets and Hierarchies: Analysis and Antitrust Implications: A Study in the Economics of Internal Organization. University of Illinois at Urbana-Champaign's Academy for Entrepreneurial Leadership Historical Research Reference in Entrepreneurship.

89. Williamson, O., 1985. The Economic Institutions of Capitalism. Free Press, New York.

90. Williamson, O., 1994. Transaction costs economics and organization theory. In: Smelser, N.J., Swedberg, R. (Eds.), The Handbook of Economic Sociology. Princeton University Press and Russell Sage Foundation, Princeton, NJ, pp. 77---107.

91. Williamson, O., 1998. The institutions of governance. The American Economic Review 88 (2), 75---79.

92. Yu, T., Subramaniam, M., Canella, A.A., 2009. Rivalry deterrence in international markets: contingencies governing the mutual forbearance hypothesis. Academy of Management Journal 52 (1), 127---147.

93. Zhu, Y., Wittman, X., Peng, M.W., 2012. Institution-based barriers to innovation in SMEs in China. Asia Pacific Journal of Management 29 (4), 1131---1142.

CHAPTER 12

INVESTIGATING THE RELATIONSHIP BETWEEN CORPORATE SOCIAL RESPONSIBILITY AND EARNINGS MANAGEMENT: EVIDENCE FROM SPAIN

Ester Gras-Gil, Mercedes Palacios Manzano, Joaquín Hernández Fernández

Department of Accounting and Finance, Faculty of Economics and Business, University of Murcia, Spain

ABSTRACT

This article investigates the relationship between corporate social responsibility and earnings management. Using panel data methodology for a sample of Spanish non-financial companies between 2005 and 2012, we find a negative impact of corporate social responsibility practices on earnings management. Corporate social responsibility is related to ethical and moral issues concerning corporate decision-making. Engaging in socially responsible activities not only improves stakeholder satisfaction, but also has a positive effect on corporate reputation.

The results show that corporate social responsibility practices may be an organizational device that leads to more effective use of resources, which then has a negative impact on earnings management practices.

Keywords: Corporate social responsibility, Discretionary accruals, Earnings management

INTRODUCTION

Earnings management has recently received considerable attention both from regulators and the popular press. Earnings management can be defined as the alteration of firms' reported economics performance by insiders to either mislead some stakeholders or to influence contractual outcomes (Healy and Wahlen, 1999; Leuz et al., 2003). Managers may be inclined to manage earnings due to the existence of explicit and implicit contracts, the firm's relation with capital markets, the need for external financing, the political and regulatory environment or several other specific circumstances (Vander Bauwhede, 2001). These deliberate managerial actions, contrived to disguise the real value of a firm's assets, transactions, or financial position, have negative consequences for shareholders, employees, the communities in which firms work, society at large, and managers' reputations, job security and careers (Zahra et al., 2005).

Accounting earnings are more reliable and more informative when managers' opportunistic behavior is controlled through a variety of monitoring systems (Dechow et al., 1996). After several recent financial scandals, there has been an international trend toward developing and implementing corporate governance mechanisms to fight against the opportunistic behaviors that have undermined investors' credibility in financial information. Corporate governance attributes help investors by aligning the interests of managers with the interests of shareholders and by enhancing the reliability of financial information and the integrity of the financial reporting process (Watts and Zimmerman, 1986).

Decades of empirical research have focused on the factors influencing the quality of earnings, specifically the accruals. However, there is also increasing attention being paid to the managerial activities which can lead to the manipulation of earnings. Previous studies document mixed results regarding the association between corporate social responsibility (CSR) and transparent financial reporting. In this article, we examine the relationship between CSR and earnings quality by using Spanish firms from 2005 to 2012. The earnings quality is measured by using the absolute value of abnormal discretionary accruals from the modified Jones model.

CSR is related to ethical and moral aspects about corporate decision-making and behavior and, as such, addresses complex issues like environmental protection, human resources management, health and safety at work, local community relations, and relationships with suppliers and customers (Castelo and Lima, 2006). Engaging in socially responsible activities not only improves stakeholder satisfaction, but also has a positive effect on corporate reputation (Orlitzky et al., 2003) and reduces the financial risk incurred by the firm (Orlitzky and Benjamin, 2001).

CSR research has employed a variety of theories and methodologies to study the potential relationship between CSR activities and other traditional measures of a firm's success (Mahoney and Roberts, 2007, p. 234). Previous studies focus on the link between CSR and economic or financial firm's performance (Moore, 2001; Orlitzky et al., 2003; Brammer et al., 2007) and the evidence is mixed. As Jorgensen and Knudsen (2006) note, this relationship represents the most questioned area of CSR (Angelidis et al., 2008); while a lot of research points in favor of a mild positive relationship (Aupperle et al., 1985; McGuire et al., 1988; Orlitzky et al., 2003; Maron, 2006; Wu, 2006; Rodgers et al., 2013) this connection has not been fully established (Neville et al., 2005; Prado-Lorenzo et al., 2008; Park and Lee, 2009) and the mechanisms through which financial performance is enhanced by CSR is not well understood (Jawahar and McLaughlin, 2001; Doh et al., 2009). The literature review suggests there remains a lack of understanding about how CSR initiatives can influence on accounting quality by reducing earnings management. Accordingly, our study fills this gap by studying the effects of CSR practices on discretionary accruals.

The country›s legal system, economic development, the importance of stock markets and ownership concentration all affect the country›s accounting standards, which in turn affect the country›s quality of financial reporting. We use Spanish data because they generally reflect an institutional setting similar to most continental countries, classified by La Porta et al. (1997) as French-origin civil law countries, where high concentration of ownership, weak investor rights and boards which are not independent of controlling shareholders are prevalent.[1] Fundamental stakeholders in the Spanish corporations include banks and industrial firms, although the role of financial institutions is not as prevalent as in other countries such as Germany and Japan, and the main agency problem arises from controlling and minority shareholders, as occurs in most European countries. The international accounting literature provides evidence that the magnitude of earnings management is on average higher in code-law countries with low investor protection rights, compared to common-law countries with high investor protection rights (Ali and Hwang, 2000; Ball et al., 2000; Leuz et al., 2003).

The publication of new European rules or codes of best practice for publicly listed companies has further intensified the discussion in Spain. However, Spain is a country in which corporate social responsibility ranks significantly behind others such as the USA and the UK. The reason must be sought in a culture which, to date, has not given sufficient importance to responsible corporate behavior. The Spanish Accounting Standards Board obliges Spanish firms to produce and disclose an environmental report in their annual accounts owing to the financial repercussions this information has on firms and the growing demand for this information from users. But social information is considered voluntary information.

The study is organized as follows. We review earlier research and provide the rationales for our hypothesis in 'Literature review and hypothesis development' section. This is followed by a presentation of the methodology, including how the sample was obtained and the data gathered and how the variables used were measured in 'Research design' section. We describe the analysis of the results in 'Results' section and the study ends with the main conclusions in 'Conclusion' section.

LITERATURE REVIEW AND HYPOTHESIS DEVELOPMENT

The impact of CSR on economic performance has received considerable attention in the literature over the past three decades (Waddock and Graves, 1997; Griffin and Mahon, 1997; McGuire et al., 1988; McWilliams and Siegel, 2000, 2001; Hillman and Keim, 2001; Simpson and Kohers, 2002; Orlitzky et al., 2003; Coombs and Gilley, 2005; Brine et al., 2006; Margolis et al., 2009; Aras et al., 2010). The instrumental stakeholder theory (Donaldson and Preston, 1995; Donaldson, 1999; Jones and Wicks, 1999) argues that good management implies positive relationships with key stakeholders, which, in turn, improve financial performance (Freeman, 1984; Waddock and Graves, 1997). The main idea here is that everything else being equal, firms that practice stakeholder management will perform better in profitability, stability and growth (Pesqueux and Damak-Ayadi, 2005).

For the purpose of this research, stakeholder theory is the accepted paradigm to explain why companies involve themselves in socially responsible activity as a strategy to maximize their long-term return on investment – sustainable business success, by recognizing the importance of each stakeholder group and incorporating this knowledge into their corporate strategy. Because the stakeholders control resources that are essential for the existence of the organization, a manager who wishes for the continued success of the firm has to strategically devote his attention to the needs of stakeholders. From this perspective, socially responsible firms are inclined to foster long-term relationships with stakeholders rather than maximize their short-term profit. The basic assumption behind this theory is that CSR may be an organizational device that leads to more effective use of resources (Orlitzky et al., 2003), which then has a positive impact on corporate financial performance. Transparent financial reporting indicates that information provided to a firm's stakeholders is more relevant to their decision making. In this regard, providing quality earnings is closely connected to CSR activities, especially in the aim to meet the needs of the stakeholders (Choi et al., 2013).

Previous research (Gelb and Strawser, 2001; Chih et al., 2008; Choi and Pae, 2001) argues that socially responsible firms are focused not only on increasing current profits but also on fostering future relationships with stakeholders (the long-term perspective hypothesis). From this perspective, socially responsible firms are inclined to foster long-term relationships with stakeholders rather than maximize their short-term profit (Choi et al., 2013). Because the stakeholders control resources that are essential for the existence of the organization, a manager who wishes for the continued success of the firm has to strategically devote his attention to the needs of stakeholders (Choi et al., 2013). In support of this view, empirical evidence demonstrates that firms that are more committed to CSR engage less in earnings management (Chih et al., 2008; Hong and Andersen, 2011; Choi et al., 2013).

In support of the long-term perspective hypothesis, the study of Chih et al. (2008) investigates whether the CSR-related features of 1.653 corporations in 46 countries had a positive or negative effect on the quality of their publicly released financial information during the 1993–2002 period. They argue that since it has neither been documented, nor globally tested whether CSR mitigates or increases the extent of earnings management, they studied three kinds of earnings management: earnings smoothing, earnings aggressiveness, and earnings losses and decreases avoidance. They find that the type of relationship between CSR and earnings management depends on which earnings management they consider. They concluded that with a greater commitment to CSR, the extent of earnings smoothing is mitigated, that of earnings losses and decreases avoidance is reduced, but the extent of earnings aggressiveness is increased. So, a firm with CSR in mind tends not to smooth earnings, and displays less interest in avoiding earnings losses and decreases. Hong and Andersen (2011) study the relationship between CSR an earnings management. They address two forms of earnings management: accruals based and activity based. Their sample consists of non-financial U.S. firms from the time period 1995–2005. They find evidence that firms which engage in CSR are less likely to manage earnings.

The paper of Choi et al. (2013) focuses on how CSR ratings are associated with earnings quality for firms with different ownership structures. They use a sample of Korean firms from 2002 to 2008. They find that CSR ratings are negatively correlated with the level of earnings management when all firms are considered. However, the relationship is weaker for chaebol firms and firms with highly concentrated ownership, which suggests that CSR practices can be abusively used by those firms to conceal their poor earnings quality.

However, an opposite view posited by some researchers suggests that CSR can be abused by managers to mask their opportunistic behaviors (the managerial opportunism hypothesis). CSR policies can be used by managers as their entrenchment strategy. Consistent with this argument, Prior et al. (2008) document a positive impact of earnings management on CSR by using archival data from a multi-national panel sample of 593 firms from 26 countries between 2002 and 2004. They argue that earnings management practices damage the collective interests of stakeholders; hence managers who manipulate earnings can deal with stakeholder activism and vigilance by resorting to CSR practices. They also suggest that executives with incentives to manage earnings will be very proactive in boosting their public exposure through CSR activities, particularly in firms with high visibility. Alternatively, firms with low levels of earnings management have fewer incentives to seek public exposure by promoting socially responsible activities.

Following the long-term perspective hypothesis, Chih et al. (2008) affirm that financial transparency and accountability (both as vital to shareholders as they are to employees, customers, communities, and leaders at all levels of society) are fast becoming principles of CSR that could reduce the extent to which insiders abuse their information advantage over outsiders. So, socially responsible firms do not undertake earnings management. The reason underlying this is that a firm that is socially responsible does not hide unfavorable earnings realizations and, therefore, conducts no earnings management (Chih et al., 2008). In this regard, Shleifer (2004) interprets that earnings manipulation, which many people find ethically objectionable, occurs less often in corporations with a strong commitment to social responsibility. CSR augments transparency and reduces the number of opportunities to manage (Chih et al., 2008). Similarly, Gelb and Strawser (2001) find that a limited sample of U.S. firms that engage in socially responsive activities provides more informative and/or extensive disclosures compared with companies that are less focused on advancing social goals. By equal measure, Shen and Chih (2005) find that greater transparency in accounting disclosure in the banking industry can reduce bank's incentive to manage earnings.

In line with previous research (Gelb and Strawser, 2001; Chih et al., 2008; Choi and Pae, 2001), we expect that socially responsible firms have low incentive to manage earnings. Alternatively, firms with poor CSR activities have more levels of earnings management. These arguments lead to the following hypothesis:

H1. The extent of corporate social responsibility is negatively associated with the level of earnings management.

RESEARCH DESIGN

Sample

The sample employed in this study consists of the 100 most reputable Spanish companies according to the Merco index for the period 2005–2012. We complement these data on corporate responsibility with financial data from SABI database. Following previous literature on earnings management, we do not consider financial firms in our analysis (Prior et al., 2008; Hong and Andersen, 2011). The final sample is an incomplete panel data of 286 firm-year observations.

Panel A of Table 1 shows the distribution of the sample by year. Panel B of Table 1provides industry breakdown.

Table 1: Sample characteristics.

	Number of firm-year observations	**Percentage of firm-year observations**
	286	
Panel A: Composition by year		
2005	46	16.08
2006	51	17.83
2007	45	15.73
2008	31	10.84
2009	25	8.75
2010	17	5.94
2011	24	8.39
2012	47	16.44

Panel B: Composition by industry		
Manufacturing	32	11.19
Construction	35	12.24
Commercial	140	48.95
Service	79	27.62
Total	286	100

Model specification

The purpose of this study is to investigate empirically how the corporate social responsibility in Spain influences the level of earnings management. Following previous studies (Warfield et al., 1995; Gabrielsen et al., 2002) we employ the absolute value of discretionary accruals [Abs(DACC)] as our measure of earnings management. In order to evaluate the effect of CSR on discretionary accruals, we regress the absolute value of discretionary accruals [Abs(DACC)] on CSR and control variables (Model 1).

$$\begin{aligned}\text{Abs(DACC)}_{it} &= \beta_0 + \beta_1\text{CSR}_{it} + \beta_2\text{LEV}_{it} + \beta_3\text{SIZE}_{it} + \beta_4\text{ROA}_{it} \\ &\quad + \beta_5\text{LIST}_{it} + \beta_6\text{BI} + \varepsilon_{it} \qquad (1)\end{aligned}$$

where Abs(DACC)_{it} is the absolute value of discretionary accruals in year *t*, scaled by lagged total assets; CSR is a is the natural logarithm of MERCO index in year *t*; LEV as end-of-year total liabilities divided by end-of-year total equity; SIZE_{it} is the natural logarithm of total assets in year *t*; ROA_{it} as net income by end-of-year total assets; LIST is a dummy variable (company listed = 1, else = 0); BI is a vector of industry dummies (Manufacturing industry, Construction industry, Commercial industry, Services).

Measurement of earnings management

Prior studies used discretionary accruals to measure the extent of earnings management. Discretionary (abnormal) accruals are defined as total accruals minus estimated normal (non-discretionary) accruals, where the estimated normal accruals can be derived from a number of discretionary accruals models widely used in prior studies (Dechow et al., 1995, 2003; Larcker and Richardson, 2004; Kotari et al., 2005; Jones et al., 2008). Since earnings management can involve either income-increasing accruals or income-decreasing accruals to meet earnings targets, consistent with prior studies (Warfield et al., 1995; Reynolds and Francis, 2000; Klein, 2002; Van Tendeloo and Vanstraelen, 2005; Bowen et al., 2008), the magnitude of absolute discretionary accruals is used in this study to assess the extent of earnings management. A higher magnitude of absolute discretionary accruals corresponds to a greater level of earnings management, or lower accounting quality, and vice versa.

Following Dechow et al. (1995), we compute the accrual component of earnings as:

$$\text{Total Accruals}_{it} = (\Delta CA_{it} - \Delta Cash_{it}) - (\Delta CL_{it} - \Delta STD_{it}) - Dep_{it} \tag{2}$$

where ΔCA_{it} = change in total current assets; $\Delta Cash_{it}$ = change in cash and cash equivalents; ΔCL_{it} = change in total current liabilities; ΔSTD_{it} = change in long-term debt included in current liabilities; Dep_{it} = depreciation and amortization expenses.

We use the cross-sectional version of the modified Jones (1991) model to estimate the non-discretionary component of total accruals (TAC) (DeFond and Jiambalvo, 1994; Yeo et al., 2002; Larcker and Richardson, 2004).

$$\frac{TAC_{it}}{A_{i,t-1}} = \beta_0 + \beta_1 \frac{\Delta REV_{it}}{A_{i,t-1}} + \beta_2 \frac{PPE_{it}}{A_{i,t-1}} + \varepsilon_{it} \tag{3}$$

For each year and industry we regress total accruals (TAC) on the change in revenues (ΔREV) and the level of gross property, plant and equipment (PPE), scaled by lagged total assets (A_{t-1}) in order to avoid problems of heteroscedasticity. The industry classification is based on the Spanish National Classification of Economic Activities (CNAE) made by the National Institute of Statistics (INE). The estimation of the regression coefficients is carried out using all Spanish firms available in the SABI database. Industry-years with fewer than six observations are excluded from the analysis (DeFond and Jiambalvo, 1994; Park and Shin, 2004).

Using the estimates for the regression parameters, $(\hat{\beta}_0, \hat{\beta}_1, \hat{\beta}_2)$, we estimate each sample firm's non-discretionary accruals (NDCA) by adjusting the change in sales for the change in accounts receivable (ΔAR) to allow for the possibility that firms could have manipulated sales by changing credit terms (Dechow et al., 1995).

$$\mathrm{NDCA}_{it} = \hat{\beta}_0 + \hat{\beta}_1 \frac{\Delta \mathrm{REV}_{it} - \Delta \mathrm{AR}_{it}}{A_{i,t-1}} + \hat{\beta}_2 \frac{\mathrm{PPE}_{it}}{A_{i,t-1}} \tag{4}$$

And we define discretionary accruals (DACC_{it}) for firm *i* in year *t* as the remaining portion of total accruals:

$$\mathrm{DACC}_{it} = \frac{\mathrm{TAC}_{it}}{A_{i,t-1}} - \mathrm{NDCA}_{it}$$

Following previous studies (Warfield et al., 1995; Gabrielsen et al., 2002) we employ the absolute value of discretionary accruals [Abs (DACC)] as our measure of earnings management. Table 2 shows the descriptive statistics of estimated discretionary accruals, and *t*-test to examine if the mean discretionary accruals are different from zero did not find evidence of income increasing or decreasing manipulation in our sample, except for the year 2010.

Table 2: Descriptive statistics of estimated discretionary accruals.

	Mean	**Median**	**SD**	***T***	**Sign**
DACC-2005	0.010	−0.003	0.144	0.47	0.63
DACC-2006	0.020	0.000	0.130	1.09	0.27
DACC-2007	0.031	0.033	0.150	1.38	0.17
DACC-2008	−0.002	−0.012	0.123	−0.09	0.92
DACC-2009	−0.003	0.003	0.104	−0.14	0.88
DACC-2010	−0.030	−0.021	0.067	−1.84	0.08*
DACC-2011	−0.009	0.001	0.088	−0.50	0.62
DACC-2012	−0.008	−0.006	0.105	−0.52	0.60
DACC-Total	0.005	−0.002	0.123	0.68	0.49

*Significantly different from zero at the 0.10 level.

DAC: Discretionary accruals using the modified Jones model.

Independent variable

We use the MERCO index to measure the extent of CSR. MERCO is the Spanish Monitor of Corporate Reputation, a tool that is already a reference for large companies in the assessment and management of their reputation, as it is the only Spanish monitor to annually evaluate (since 2000) the reputation of the companies that operate in Spain, as is done by Fortune or The Financial Times.

Merco seeks to measure and assess the various facets that make up a firm's reputation. It uses a process based on a variety of steps that are aimed at collecting data from different sources of information. The processes used by Merco to obtain the information required are:

1. Interviews with directors. The aim of this interview is to ascertain the opinion held by directors of the most important companies in Spain about the corporate reputation of firms operating in our country. Within the appraisals of these agents, the Merco questionnaire distinguishes between two areas: overall rating, in which a director rates up to 10 companies which he or she considers to enjoy the best reputation in his or her country, regardless of the firms' activities; sectorial rating, in which those answering the questionnaire are asked to indicate the two firms in their own sector that enjoy the best reputation.
2. Assessment by experts. This seeks to incorporate in the corporate reputation rating the views of several outside agents. In this stage, the 90 firms selected in the previous process are rated by 5 groups of experts: Financial Analysts; Consumer Associations; NGOs; Trade Unions; and Economic Journalists (Table 3).

Table 3: Variables assessment by group of experts.

Group of experts	Merco dimension	Rated variable
Financial Analysts	Economic-financial results	Economic-financial results Quality of the economic information.
NGO	Corporate ethics and responsibility	Commitment with the community. Social and environmental responsibility.
Consumer Associations	Quality of the commercial offer	Quality of product – service. Respect for consumer rights.
Trade Unions	Internal reputation	Labor quality.
Economic Journalists	Corporate ethics and responsibility	Transparency of information. Accessibility.

Own: Merco (2013).

3. Direct Assessment. The aim is to have a rating by qualified specialists in Analysis and Research of the relative merits and corporate reputation presented by the 90 firms selected from the provisional ranking. This rating will require firms to accredit their reputational values via a questionnaire and documents that back up the information in the same. The variables analyzed by the specialists are shown in Table 4.

Table 4: Variables analyzed by the specialists.

Merco dimension	Rated variable
Economic and financial results	Sheetprofit. Profitability. Quality of the economic information.
Quality of the commercial offer	Product values. Brand values. Customer service and care.
Internal reputation	Labor quality. Ethical and professional values. Identification with the business project.
Corporate ethics and responsibility	Ethical corporate behavior. Commitment to the community. Social and environmental responsibility.
International dimension of the firm	Number of countries in which it operates. Turnover abroad. International strategic alliances.
Innovation	Investments in R+D. New products and services. New channels.

Own: Merco (2013).

4. Merco Tracking. This phase seeks to obtain a rating from the population as a whole for the reputation of the firms selected and to detect any possible evolution of the ranking.

Merco Persons is a monitor that seeks to identify the attractiveness of the various firms in terms of places to work in. The starting point here is the views of various agents: the firm's employees; university students; former students of business schools; the general population; and human resource managers. It also uses benchmarking to make a comparison of the main indicators of people management.

The final ranking is calculated at the close of each phases outlined as the weighted sum of the scores obtained in those phases. The weighting used are given in Table 5.

Table 5: Weighting percentages.

Overall ranking	37%
Sectoral ranking	8%
Financial analysts	8%
Trade unions	5%
Consumer associations	5%
NGOs	5%
Economic journalists	8%
Merco tracking	8%
Merco persons	8%
Direct assessment	8%

Own: Merco (2013).

Firms which do not achieve at least 8 points from other firms in the overall ranking or that receive all their overall points from a single participating firm, aside from their own, are excluded from the final ranking. If necessary, the general ranking components are given a weighting to remove the corporate vote effect.

Thus, after totaling all the Merco components and correcting the possible undesired effects (group votes, massive votes for organizations of small importance, etc.), we obtain a score index over 10,000 and the ranking derived from it, which is published annually.

Control variables

To control for differences in earnings management incentives, we include the following variables based on prior research (Ashbaugh, 2001; Pagano et al., 2002; Tarca, 2004; Barth et al., 2006; Lang et al., 2006). First, we include the natural logarithm of total assets (SIZE) to proxy for the size of a company. The political cost hypothesis states that larger firms are more likely to prefer downward earnings management, because the potential for government scrutiny increase as firms are larger and more profitable (Watts and Zimmerman, 1990; Young, 1999). Closer scrutiny by outsiders can potentially reduce managers' opportunities to exercise their accounting discretion in big firms. The expected relationship between discretionary accruals and firm size is negative.

Second, the variable leverage (LEV) controls for the likelihood of bankruptcy. A higher total debt to asset ratio indicates a higher possibility of debt covenant violation, which creates an incentive to increase reported earnings thorough accruals-based earnings management. According to Park and Shin (2004), firms that face financial constraints or distress have an incentive to adjust earnings upward in order to avoid a potential loss from disclosing a financial problem. These arguments would predict a positive relationship between discretionary accruals and financial leverage. Press and Weintrop (1990) and Sweeney (1994) also report that firms respond to debt contracting by strategically reporting discretionary accruals.

Third, the return on assets (ROA) is included as a proxy for the profitability of a firm. Dechow et al. (1995) and McNichols (2000) report that firms with abnormally high (low) earnings have positive (negative) shocks to earnings that include an accrual component and thus, firms with high (low) earnings tend to have high (low) accruals. As a consequence, one is more likely to find a positive relationship for the most profitable firms.

Finally, Lang et al. (2003) find that cross-listed firms appear to be less aggressive in terms of earnings management and report accounting data that are more conservative. Firms with a foreign exchange listing are presumed to have greater incentives to report transparently because they are subject to restrictions imposed by different countries and are exposed to a higher litigation risk. Therefore, we include a dummy variable to control if is a cross-listed firm or not. It can be expected that earnings quality is enhanced when listed on an international capital market (Ball et al., 2000, 2003).

We also include industry dummies to control for industry effects on earnings management.

Following Petersen (2009), we use t-statistics based on standard errors clustered at the firm and the year level, which are robust both to heteroscedasticity and within-firm serial correlation.[2]

RESULTS

Descriptive statistics

The descriptive statistics of the variables are shown in Table 6. The mean value of absolute earnings management moves around 0.089. This value is higher in our sample in comparison to that of Choi et al. (2013) and Prior et al. (2008). The descriptive analysis shows that the CSR variable (natural logarithm of Merco index) shows a mean value of 7.90.

Table 6: Descriptive statistics.

Variable	Mean	Median	STD	Min	Max
Abs (DACC)	0.089	0.061	0.085	0.0003	0.3668
CSR	7.904	7.908	0.534	7.071	9.210
LEV	0.645	0.664	0.203	0.118	1.512
SIZE	12.234	12.000	2.187	6	18
ROA	0.034	0.033	0.063	−0.454	0.390
	Yes (%)	**No (%)**			
LIST	74.2	25.8			

Where Abs(DACC) is the absolute value of discretionary accruals using Jones modified model; CSR: natural logarithm of Merco index in year *t*; LEV: as end-of-year total liabilities divided by end-of-year total assets; SIZE: natural logarithm of total assets in year *t*; ROA: as net income divided by end-of-year total assets; LIST: dummy variable (company listed = 1, else = 0).

In Table 7, the absolute value of discretionary accruals is displayed by CSR intervals, suggesting a negative relation between CSR and discretionary accruals with an inflection point for low levels of CSR.

Table 7: Discretionary accruals by CSR.

Range of CSR	Mean Abs(DACC)
<10%	0.095
10–25%	0.092
25–50%	0.073
50–75%	0.085
>75%	0.075

Abs(DACC): absolute value of discretionay accruals using the modified Jones model.

The correlation matrix between variables is presented in Table 8. The correlation coefficient of CSR with size is significantly positive. The correlation coefficient of EM with leverage and listing firms are significantly positive, and significantly negative with size.

Table 8: Correlation matrix.

	Abs(DACC)	CSR	LEV	SIZE	ROA	LIST
Abs (DACC)	1					
CSR	−0.056	1				
LEV	0.133*	−0.095	1			
SIZE	−0.153**	0.154**	0.062	1		
ROA	0.015	−0.048	−0.273**	0.056	1	
LIST	0.172**	−0.045	0.018	−0.608**	−0.058	1

*Significantly different from zero at the 0.05 level.

**Significantly different from zero at the 0.01 level.

zWhere Abs(DACC) is the absolute value of discretionary accruals using Jones modified model; CSR: natural logarithm of Merco in year *t*; LEV: as end-of-year total liabilities divided by end-of-year total assets; SIZE: natural logarithm of total assets in year *t*; ROA: as net income divided by end-of-year total assets; LIST: as a dummy variable taking the value 1 if it is a listing firm and 0 if it is a no-listing firm.

Multivariate analysis

The results of Model 1 are presented in Table 9. We use *t*-statistics based on standard errors clustered at the firm and the year level (Petersen, 2009), which are robust both to heteroscedasticity and within-firm serial correlation. The results show a consistently significant negative relationship between corporate social responsibility practiccs (CSR) and absolute discretionary accruals. The discretionary accruals are substantially lower to more socially responsible firms, and this difference is statistically significant at the 0.05 level. These results provide strong evidence for the effect of corporate social responsibility practices in reducing earnings management by Spanish firms. Our findings provide support for Hypothesis 1.

Table 9: Regressions of absolute discretionary accruals on independent variables and control variables.

Model: $\text{Abs(DACC)}_{it} = \beta_0 + \beta_1\text{CSR}_{it} + \beta_2\text{LEV}_t + \beta_3\text{SIZE}_{it} + \beta_4\text{ROA}_{it} + \beta_5\text{ LIST}_{it} + \varepsilon_{it}$		
	Model	
Variables	**Estimated coefficient**	***t*-Statistic**
Intercept	0.1949	4.71***
CSR	−0.012	−1.93**
LEV	0.0458	2.15**
SIZE	−0.0040	−1.72*
ROA	0.1348	2.29**
LIST	0.0064	0.75
Industry dummies	Yes	
N	286	
R^2 **(adjusted)**	0.132	
F	5.28***	

*Significantly different from zero at the 0.10 level (two-tailed).

**Significantly different from zero at the 0.05 level (two-tailed).

***Significantly different from zero at the 0.01 level (two-tailed).

Where Abs(DACC) is the absolute value of discretionary accruals using Jones modified model; CSR: natural logarithm of Merco index in year *t*; LEV: as end-of-year total liabilities divided by end-of-year total assets; SIZE: natural logarithm of total assets in year *t*; ROA: as net income divided by end-of-year total assets; LIST: dummy variable (company listed = 1, else = 0). Models include industry dummies. Regressions are run using two-way cluster standard errors (Petersen, 2009) at the time and firm level which are robust to both heteroscedasticity and within-firm serial correlation.

These results are consistent with the findings of Chih et al. (2008), who argue that companies with greater CSR conduct less earning smoothing. However, these results are in contrast with the findings of Prior et al. (2008), who found a positive impact of earnings management practices on CSR. They explain such a result by the fact that managers who indulge in earnings management practices have two reasons to satisfy stakeholders' interest. First, there is a pre-emptive reason-managers anticipate that stakeholder activism, as a result of earnings manipulation, may damage their position in the firm. A good way of avoiding such activism is to satisfy stakeholders' interests. Second, an entrenchment reason-manager tends to collude with other stakeholders as a hedging strategy against disciplinary initiatives from shareholders affected detrimentally by these earnings management practices.

In terms of the control variables, we find that absolute discretionary accruals are decreasing in size, but increasing in leverage and profitability. Agency and political costs explain why firm size is significantly associated with earnings management. The coefficient size is significant and negative. Firm size is clearly a business characteristic that exhibits differences in the degree of earnings management, showing that high size firms have less discretionary accruals. This is often used as a proxy for political sensitivity. Large firms with large profits may try to manage earnings downwards (Zimmerman, 1983; Liberty and Zimmerman, 1986). The larger the firm, the more likely managers are to choose income-decreasing accruals.

The control variable Leverage (Lev) is found to be significantly and positively related to absolute discretionary accruals. High leverage has been found to be associated with closeness to the violation of debt covenants (Press and Weintrop, 1990), and debt covenant violation has been found to be associated with discretionary accrual choice (DeFond and Jiambalvo, 1994). To avoid debt covenant violation, managers of highly leveraged firms have incentives to make income-increasing discretionary accruals. However, high leverage is also associated with financial distress (Beneish and Press, 1995). According to DeAngelo et al. (1994), troubled companies have large negative accruals related to contractual renegotiations that provide incentives to reduce earnings. Our results are consistent with Watts and Zimmerman (1978), DeFond and Jiambalvo (1994), Sweeney (1994) and Dichev and Skinner (2002).

Robustness test

To examine the robustness of our results, we carry out an additional analysis. Different measures of discretionary accruals are used in the literature to proxy for earnings management. There exists extensive literature relating discretional accruals and free cash flow (Chung et al., 2005; Dechow and Ge, 2006; Bukit and Iskandar, 2009). These studies establish that the opportunity for earnings management is higher among companies with high surplus free cash flow. Free cash flow (FCF) is defined as the excess of cash available to a firm after it has invested in all positive-net-present-value projects that is not paid out in dividends (Jensen, 1986). Managers of low growth firms with high FCF are expected to be involved in non-value maximizing activities including an increase in perquisite consumption and compensation at the expense of shareholders as well as the manipulation of accounting number (Jensen, 1989; Shleifer and Visny, 1989; Lang et al., 1991; Christie and Zimmerman, 1994).

We therefore rerun our model but using like dependent variable the free cash flow (Model 2). Free cash flow is measured by operating income before depreciation minus expenses such as tax expense, interest expense and dividend (Lehn and Poulsen, 1989).

$$\mathsf{FCF}_{it} = \beta_0 + \beta_1\,\mathsf{CSR}_{it} + \beta_2\,\mathsf{LEV}_t + \beta_3\,\mathsf{SIZE}_{it} + \beta_4\,\mathsf{ROA}_{it} + \beta_5\,\mathsf{LIST}_{it} + \varepsilon_{it}$$

The results of this analysis are shown in Table 10. Ours results indicate that CSR has a negative impact on FCF. The results can be interpreted that minor CSR high free cash flow. These results are consistent with our Model 1, as managers of high free cash flow companies have incentive to engage in earnings management. The results again support Hypothesis 1. In term of control variables, we find that free cash flow is increasing in size and in ROA. The coefficients are significant and positive.

Table 10: Regressions of cash flow on independent variables and control variables.

Variables	**Estimated coefficient**	***t*-Statistic**
Intercept	4.650	5.78***
CSR	−0.374	−2.48**
LEV	0.093	0.18
SIZE	0.964	16.77***
ROA	6.108	3.98***
LIST	−0.143	−0.47
Industry dummies	Yes	
N	286	
R^2 (adjusted)	0.819	
F	123.73***	

**Significantly different from zero at the 0.05 level (two-tailed).

***Significantly different from zero at the 0.01 level (two-tailed).

Where FCF: natural logarithm of free cash flow in year *t*; CSR: natural logarithm of Merco index in year *t*; LEV: as end-of-year total liabilities divided by end-of-year total assets; SIZE: natural logarithm of total assets in year *t*; ROA: as net income divided by end-of-year total assets; LIST: dummy variable (company listed = 1, else = 0). Models include industry dummies. Regressions are run using two-way cluster standard errors (Petersen, 2009) at the time and firm level which are robust to both heteroscedasticity and within-firm serial correlation.

CONCLUSIONS

The motivation for firms to engage in CSR has been an unresolved issue for which previous research has yielded mixed results. While some researchers suggest that CSR engagement is induced by the long-term perspectives for sustainable operations of a business, others argue that CSR is a practice used by managers who are involved in opportunistic behaviors. Stakeholder theory is the accepted paradigm to explain why companies involve themselves in socially responsible activity as a strategy to maximize their long-term return on investment – sustainable business success, by recognizing the importance of each stakeholder group and incorporating this knowledge into their corporate strategy. Stakeholder theory regards CSR as a success medium for managing stakeholder relationships (Ullman, 1985; Roberts, 1992).

In this article, we examine the relationship between CSR and earnings quality by using Spanish firms from 2005 to 2012. For the analysis, the MERCO index is employed as a proxy for the CSR ratings of Spanish firms. The earnings quality is measured by using the absolute value of abnormal discretionary accruals from the modified Jones model. In particular, we explore the thesis that managers manipulate earnings in order to obtain private benefits, and through these practices they damage the interests of stakeholders. A firm with CSR in mind tends not to manipulate earnings because it is not a responsible practice. And such firms, therefore, should act in a responsible manner when reporting accounting information.

The empirical results confirm to our theoretical contention. In particular, we find a negative impact of CSR practices on earnings management, so firms that are more committed to CSR engage less in earnings management. These results are consistent with Chih et al. (2008). They find that companies with higher social responsibility engage in less earnings smoothing and less earnings decrease/loss avoidance. Similarly, Shleifer (2004) interprets that earnings manipulation, which many people find ethically objectionable, occurs less often in corporations with a strong commitment to social responsibility. As suggested by Hong and Andersen (2011), more socially responsible firms have higher quality accruals and less activity-based earnings management, both of which impact financial reporting quality.

Socially responsible firms are inclined to foster long-term relationships with stakeholders rather than maximize their short-term profit. In this regard, providing quality earnings is closely connected to CSR activities, especially in that both aim to meet the needs of the stakeholders. This reason to engage in social responsibility is supported and supplemented by the log stream of empirical research that finds that CSR and financial performance are positively related (Orlitzky et al., 2003; Brammer et al., 2007). In this regard, as Labelle et al. (2010) indicate, higher level of corporate moral development is associated with higher quality financial reporting.

Analysis of the determinants of accounting quality has important policy implications. Since all EU countries will have consistent financial reporting rules, future improvements in accounting quality will be largely dependent on changes in a country›s legal and political system and financial reporting incentives. Changing a country›s overall institutional infrastructure is difficult, so addressing financial reporting incentives will perhaps be the least costly means of achieving any further improvements in accounting quality. This study offers insights for policy makers and managers interested in enhancing CSR.

These results must be interpreted with some limitations in mind. The study is not free from external validity problems caused by a restricted sampling frame and small sample size. Due to the feasibility of collecting CSR and financial data, the study includes in the sample only the most reputable companies in Spain. To enhance generalizability, future research using increased numbers of observations including all Spanish companies is encouraged.

Furthermore, the results show some increase of accruals for companies with a medium-high profile of CSR activities. This may reveal a potential non-linear relationship (maybe quadratic) between CSR and earnings management. Arguments taken from the different theories also may support this idea. Thus, future research could consider this issue as an interesting line of investigation.

REFERENCES

1. Ali, A., Hwang, L., 2000. Country-specific factors related to financial reporting and the value relevance of accounting data. J. Account. Res. 38, 1---23.
2. Angelidis, J.P., Massetti, B.L., Magee-Egan, P., 2008. Does corporate social responsibility orientation vary by position in the organizational hierarchy? Rev. Bus. 28 (3), 23---32.

3. Aras, G., Aybars, A., Kutlu, O., 2010. Managing corporate performance. Investigating the relationship between corporate social responsibility and financial performance in emerging markets. Int. J. Product. Perform. Manage. 59 (3), 229---254.
4. Ashbaugh, H., 2001. Non-U.S. firms' accounting standard choices. J. Account. Public Policy 20 (2), 129---153.
5. Aupperle, K., Carroll, A., Hatfield, J., 1985. An empirical examination of the relationship between corporate social responsibility and profitability. Acad. Manage. J. 28 (2), 446---463.
6. Ball, R., Kothari, S., Robin, A., 2000. The effect of international institutional factors on properties of accounting earnings. J. Account. Econ. 29 (1), 1---51.
7. Ball, R., Robin, A., Wu, J.S., 2003. Incentives versus standards: properties of accounting income in four East Asian countries. J. Account. Econ. 36 (1---3), 235---270.
8. Barth, M.E., Landsman, W.R., Lang, M., Williams, C., 2006. Accounting Quality: International Accounting Standards and US GAAP.
9. Working Paper, Stanford University, University of North Carolina. Beneish, M.D., Press, E., 1995. The resolution of technical default. Account. Rev. 70, 337---354.
10. Brammer, S., Millington, A., Rayton, B., 2007. The contribution of corporation social responsibility to organizational commitment. Int. J. Hum. Resour. Manage. 18 (10), 1701---1719.
11. Brine, M., Brown, R., Hackett, G., 2006. Corporate Social Responsibility and Financial Performance in the Australian Context. Australian Treasury.
12. Bowen, R., Rajgopal, S., Venkatachalam, M., 2008. Accounting discretion, corporate governance, and firm performance. Contemp. Account. Res. 25 (2), 351---405.
13. Bukit, R.B., Iskandar, T.M., 2009. Surplus free cash flow, earnings management and audit committee. Int. J. Econ. Manag. 3 (1), 204---223.
14. Castelo, M., Lima, L., 2006. Corporate social responsibility and resource-based perspectives. J. Bus. Ethics 69,111---132.

15. Chih, H.L., Shen, C.H., Kang, F.C., 2008. Corporate social responsibility, investor protection, and earnings management: some international evidence. J. Bus. Ethics 79, 179---198.
16. Choi, B.B., Lee, D., Park, Y., 2013. Corporate social responsibility, corporate governance and earnings quality: evidence from Korea. Corp. Gov.: Int. Rev. 21 (5), 447---467.
17. Choi, T.H., Pae, J., 2001. Business ethics and financial reporting quality: evidence from Korea. J. Bus. Ethics 103, 403---427.
18. Christie, A.A., Zimmerman, 1994. Efficient and opportunistic choices of accounting procedures. Corporate control contest. Account. Rev. 69 (4), 539---566.
19. Chung, R., Firth, M., Kim, J.-B., 2005. Earnings management, surplus free cash flow, and external monitoring. J. Bus. Res. 58, 766---776.
20. Coombs, J.E., Gilley, K.M., 2005. Stakeholder management as a predictor of CEO compensation: main effects and interactions with financial performance. Strat. Manage. J. 26 (9), 827---841.
21. DeAngelo, H., DeAngelo, L., Skinner, D.J., 1994. Accounting choice in troubled companies. J. Account. Econ. 17 (1---2), 113---143.
22. Dechow, P., Sloan, R., Sweeney, A.P., 1995. Detecting earnings management. Account. Rev. 70 (2), 193---225.
23. Dechow, P., Sloan, R., Sweeney, A.P., 1996. Causes and consequences of earnings manipulation: an analysis of firms subject to enforcement actions by the SEC. Contemp. Account. Res. 13
24. (1), 1---36.
25. Dechow, P., Richardson, S., Tuna, I., 2003. Why are earnings kinky? An examination of the earnings management explanation. Rev. Account. Stud. 8 (2), 355---384.
26. Dechow, P.M., Ge, W., 2006. The persistence of earnings and cash flows and the role of special items: implications for the accrual anomaly. Rev. Account. Stud. 11, 253---296.
27. DeFond, M.L., Jiambalvo, J., 1994. Debt covenant violation and manipulation of accruals. J. Account. Econ. 17 (1---2), 145---176.

28. Dichev, I.D., Skinner, D.J., 2002. Large-sample evidence on the debt covenant hypothesis. J. Account. Res. 40 (4), 1091---1123.
29. Doh, J.P., Bunyaratavej, K., Hahn, E., 2009. Separable but not equal: the location determinants of discrete services offshoring activities. J. Int. Bus. Stud. 40 (6), 926---943.
30. Donaldson, T., Preston, L.E., 1995. The stakeholder theory of the corporation: concepts, evidence, and implications. Acad. Manage. Rev. 20, 701---711.
31. Donaldson, L., 1999. Making stakeholder theory whole. Acad. Manage. Rev. 24 (2), 237---241.
32. Freeman, E., 1984. Strategic Management: A Stakeholder Approach. Pitman, Boston.
33. Gabrielsen, G., Gramlich, J., Plenborg, T., 2002. Managerial ownership, information content of earnings, and discretionary accruals in a non-US setting. J. Bus. Financ. Account. 29, 967---988.
34. Gelb, D.S., Strawser, J.A., 2001. Corporate social responsibility and financial disclosure: an alternative explanation for increased disclosure. J. Bus. Ethics 33, 1---13.
35. Griffin, J.J., Mahon, J.F., 1997. The corporate social performance and corporate financial performance debate: twenty-five years of incomparable research. Bus. Soc. 36, 5---31.
36. Healy, P.M., Wahlen, J.M., 1999. A review of the earnings management literature and its implications for standard setting. Account. Horiz. 13, 365---383.
37. Hillman, A.J., Keim, G.D., 2001. Shareholder value, stakeholder management, and social issues: what's the bottom line? Strat. Manage. J. 22 (2), 125---139.
38. Hong, Y., Andersen, M.L., 2011. The relationship between corporate social responsibility and earnings management: an exploratory study. J. Bus. Ethics 104, 461---471.
39. Jawahar, I.M., McLaughlin, G.L., 2001. Toward a descriptive stakeholder theory: an organizational life cycle approach. Acad. Manage. Rev. 26, 397---414.

40. Jensen, M.C., 1986. Agency costs of free cash flow, corporate finance and takeovers. Am. Econ. Rev. 76, 323---329.
41. Jensen, M.C., 1989. Eclipse of the public corporation. Harvard Bus. Rev. 5, 61---74.
42. Jones, J., 1991. Earnings management during import relief investigations. J. Account. Res. 29 (2), 193---228.
43. Jones, T.M., Wicks, A.C., 1999. Convergent stakeholder theory. Acad. Manage. Rev. 24 (2), 206---221.
44. Jones, K., Krishnan, G., Melendrez, K., 2008. Do models of discretionary accruals detect actual cases of fraudulent and restated earnings? An empirical analysis. Contemp. Account. Res. 25 (2), 499---531.
45. Jorgensen, A., Knudsen, J.S., 2006. Sustainable competitiveness in global value chains. How do small Danish firms behave? Corp. Gov. 6 (special issue), 449---462.
46. Klein, A., 2002. Audit Committee, Board of Director characteristics, and earnings management. J. Account. Econ. 33, 375---400.
47. Kotari, S.P., Leone, A.J., Wasley, C.E., 2005. Performance matched discretionary accrual measures. J. Account. Econ. 39, 163---197. La Porta, R., Lopez de Silanes, F., Shleifer, A., Vishny, R.W., 1997. Law and finance. J. Financ. 52, 1131---1150.
48. La Porta, R., Lopez de Silanes, F., Shleifer, A., 1999. Corporate ownership around the world. J. Financ. 54, 471---517.
49. Labelle, R., Gargouri, R.M., Francoeur, C., 2010. Ethics, diversity management and financial reporting quality. J. Bus. Ethics 93, 335---353.
50. Lang, L.H.P., Stulz, R.M., Wlaking, R.A., 1991. A test of the free cash flow hypothesis, the case of bidder returns. J. Financ. Econ. 29, 315---335.
51. Lang, M., Lins, K.V., Miller, D.P., 2003. ADRs, analysts, and accuracy: does cross listing in the United States improve a firm's information environment and increase market value? J. Account. Res. 41 (2), 317---345.
52. Lang, M., Smith Raedy, J., Wilson, W., 2006. Earnings management

and cross listing: are reconciled earnings comparable to US earnings? J. Account. Econ. 42 (1---2), 255---283.

53. Larcker, D.F., Richardson, S.A., 2004. Fees paid to audit firms, accrual choices, and corporate governance. J. Account. Res. 42, 625---658.
54. Lehn, K., Poulsen, A., 1989. Free cash flow and stakeholder gains in going private transactions. J. Financ. 44 (3), 771---787.
55. Leuz, C., Nanda, D., Wysocki, P.D., 2003. Earnings management and investor protection: an international comparison. J. Financ. Econ. 69 (5), 505---527.
56. Liberty, S., Zimmerman, J.L., 1986. Labor union contract negotiations and accounting choices. Account. Rev. 61 (4), 692---712.
57. Mahoney, L., Roberts, R.W., 2007. Corporate social performance, financial performance and institutional ownership in Canadian firms. Account. Forum 31 (3), 233---253.
58. Margolis, J.D., Elfenbein, H.A., Walsh, J.P., 2009. Does it pay to be good. . .and does it matter? A meta-analysis of the relationship between corporate social and financial performance. Working paper, Harvard University, Cambridge, MA, http://dx.doi.org/10.2139/ssrn.1866371, Available at:
59. http://ssrn.com/abstract=1866371.
60. Maron, I.Y., 2006. Toward a unified theory of the CSP-CFP link. J. Bus. Ethics 67 (2), 191---200.
61. McGuire, J., Sundgren, A., Schneeweis, T., 1988. Corporate social responsibility and firm financial performance. Acad. Manage. J. 31 (4), 854---872.
62. McNichols, M., 2000. Research design issues in earnings management studies. J. Account. Public Policy 19 (4-5), 313---345.
63. McWilliams, A., Siegel, D., 2000. Corporate social responsibility and financial performance: correlation or misspecification? Strat. Manage. J. 21, 603---609.
64. McWilliams, A., Siegel, D., 2001. Corporate social responsibility: a theory of the firm perspective. Acad. Manage. Rev. 26 (1), 117---127.

65. Merco, 2013. El proceso de elaboración de Merco Espana.˜ Monitor Empresarial de Reputación Corporativa, http://www.merco. info/ datafiles/0000/4004/El proceso de elaboracion de Merco 2013.pdf.
66. Moore, G., 2001. Corporate social and financial performance: an investigation in the U.K. supermarket industry. J. Bus. Ethics 34, 299---315.
67. Neville, B.A., Bell, S.J., Mengüc,, B., 2005. Corporate reputation, stakeholders and the social performance---financial performance relationship. Eur. J. Mark. 39 (9---10), 1184---1198.
68. Orlitzky, M., Benjamin, J.D., 2001. Corporate social performance and firm risk: a meta-analysis review. Bus. Soc. 40 (4), 369---387.
69. Orlitzky, M., Schmidt, F.L., Rynes, S.L., 2003. Corporate social and financial performance: a meta-analysis. Organ. Stud. 24, 403---441.
70. Pagano, M., Röell, A., Zehner, J., 2002. The geography of equity listing: why do companies list abroad? J. Financ. 57 (6), 2651---2694.
71. Park, S., Lee, S., 2009. Financial rewards for social responsibility. Cornell Hosp. Q. 50 (2), 168---179.
72. Park, Y.W., Shin, H.H., 2004. Board composition and earnings management in Canada. J. Corp. Financ. 10, 431---457.
73. Pesqueux, Y., Damak-Ayadi, S., 2005. Stakeholder theory in perspective. Corp. Gov. 5 (2), 5---21.
74. Petersen, M.A., 2009. Estimating standard errors in finance panel data sets: comparing approaches. Rev. Financ. Stud. 22 (1), 435---480.
75. Prado-Lorenzo, J.M., Gallego-Alvarez, I., Garcia-Sanchez, I.M., Rodriguez-Dominguez, L., 2008. Social responsability in Spain. Practices and motivations in firms. Manage. Decis. 46 (8), 1247---1271.
76. Press, E., Weintrop, J., 1990. Accounting-based constraints in public and private debt agreements: their association with leverage and impact on accounting choice. J. Account. Econ. 12,65---95.
77. Prior, D., Surroca, J., Tribo, J.A., 2008. Are socially responsible managers really ethical? Exploring the relationship between

earnings management and corporate social responsibility. Corp. Gov. 16 (3), 160---176.

78. Reynolds, J.K., Francis, J., 2000. Does size matter? The influence of large clients on office-level auditor reporting decisions. J. Account. Econ. 30 (3), 375---400.
79. Roberts, R.W., 1992. Determinants of corporate social responsibility disclosure: an application of stakeholder theory. Account. Organ. Soc. 17, 595---612.
80. Rodgers, W., Choy, H.L., Guiral, A., 2013. Do investors value a firm's commitment to social activities? J. Bus. Ethics 114 (4), 607---623.
81. Shen, C.H., Chih, H.L., 2005. Investor protection, prospect theory, and earnings management: an international comparison of the banking industry. J. Bank. Financ. 29, 2675---2697.
82. Shleifer, A., Visny, R.W., 1989. Management entrenchment, the case of manager-specific investment. J. Financ. Econ. 25, 123---139.
83. Shleifer, A., 2004. Does Competition Destroy Ethical Behavior? Working Paper. Harvard University.
84. Simpson, W., Kohers, T., 2002. The link between social and financial performance: evidence from the banking industry. J. Bus. Ethics 35, 97---109.
85. Sweeney, A.P., 1994. Debt-covenant violations and managers' accounting responses. J. Account. Econ. 17 (3), 281---308.
86. Tarca, A., 2004. International convergence of accounting practices: choosing between IAS and U.S. GAAP. J. Int. Financ. Manage. Account. 15 (1), 60---91.
87. Ullman, A.E., 1985. Data in search of a theory: a critical examination of the relationships among social performance, social disclosure and economic performance of US firms. Acad. Manage. Rev. 10, 540---557.
88. Van Tendeloo, B., Vanstraelen, A., 2005. Earnings management under German GAAP versus IFRS. Eur. Account. Rev. 14 (1), 155---180.
89. Vander Bauwhede, H., 2001. Earnings Management in a Non-

90. Anglo-Saxon Environment: The Impact of Audit Quality, Public Ownership and Stakeholders. Catholic University Leuven, Leuven.
91. Waddock, S.A., Graves, S.B., 1997. The corporate social performance---financial performance link. Strat. Manage. J. 18, 303---319.
92. Warfield, T., Wild, J., Wild, K., 1995. Managerial ownership, accounting choices and informativeness of earnings. J. Account. Econ. 20, 61---91.
93. Watts, R., Zimmerman, J., 1978. Towards a positive theory of the determination of accounting standards. Account. Rev. 53 (1), 112---134.
94. Watts, R., Zimmerman, J., 1986. Positive Accounting Theory. Prentice Hall, Englewood Cliffs, New Jersey.
95. Watts, R., Zimmerman, J., 1990. Positive accounting theory: a ten year perspective. Account. Rev. 65 (1), 131---156.
96. Wu, M.L., 2006. Corporate social performance, corporate financial performance, and firm size: a meta-analysis. J. Am. Acad. Bus. 8 (1), 163---171.
97. Yeo, G.H.H., Tan, P., Ho, K.W., Chen, S.S., 2002. Corporate ownership structure and the informativeness of earnings. J. Bus. Financ. Account. 29, 1023---1046.
98. Young, S., 1999. Systematic measurement error in the estimation of discretionary accruals: an evaluation of alternative modelling procedures. J. Bus. Financ. Account. 26 (7-8), 833---862.
99. Zahra, S.A., Priem, R.L., Rasheed, A.A., 2005. The antecedents and consequences of top management fraud. J. Manage. 31, 803---828.
100. Zimmerman, J.L., 1983. Taxes and firm size. J. Account. Econ. 5(2), 119---149.

CHAPTER 13

RELATIONSHIP BETWEEN CORPORATE SOCIAL RESPONSIBILITY AND COMPETITIVE PERFORMANCE IN SPANISH SMES: EMPIRICAL EVIDENCE FROM A STAKEHOLDERS' PERSPECTIVE

Jesús Herrera Madueno˜ [a], Manuel Larrán Jorge[a], Isabel Martínez Conesa [b], Domingo Martínez-Martínez [a]

[a] Financial Economics and Accounting Department at University of Cádiz, Business and Administration College, Av. Duque de Nájera, 8, 11002 Cádiz, Spain

[b] Financial Economics and Accounting Department at University of Murcia; Business and Administration College, Facultad de Economía y Empresa, Campus de Espinardo, 30100 Murcia, Spain

ABSTRACT

In a financial economic scenario in which the corporate survival of small and medium enterprises (SMEs) is more conditioned than ever by competitive performance, this paper aims to show that the strategic incorporation of socially responsible actions, more concerned and engaged with stakeholders, contributes to improve the competitiveness of these organizations. Thus, the existence of a direct or mediated relationship between the development of Corporate Social Responsibility (CSR) practices and competitive performance has been analyzed from a multi-stakeholder perspective. To accomplish this task, data were collected from a sample of 481 Spanish SMEs and the technique of partial least squares (PLS) was used. Outcomes show that the development of CSR practices contributes to increase the competitive performance both directly and indirectly, through the ability of these organizations to manage their stakeholders. This study, therefore, supports the social impact hypothesis and offers evidence about some intangibles such as the relational capacity mediate the causal effect between CSR and competitive performance.

Keywords: CSR, Competitive performance, Relational capacity, Stakeholders, SMEs

INTRODUCTION

Corporate Social Responsibility (CSR), Social Responsibility (SR) or any of its aspects are gaining great attention in the academic and professional fields. Companies are increasingly more aware and encouraged to integrate and participate in CSR issues (Mark-Herbert and Von Schantz, 2007). It has been pointed out as an essential concept that business managers should understand and manage. Firms of all types and sizes are called to become socially responsible, ecologically sustainable and economically competitive (Orlitzky et al., 2011). However, the real development that CSR has experienced in small and medium enterprises (SMEs) is different than in large corporations.

In general, SMEs have certain characteristics that do interesting developing detailed and specific studies focused on them. This kind of organization usually shows particular strategies and structures, less formalized and more dependent on the disposition, participation, and ability to design strategies of the owners/managers, which complicate the development of CSR actions similar to large corporations' practices. The lack of knowledge that SME managers usually have about CSR, the close relations with stakeholders (Fisher et al., 2009; Russo and Tencati, 2009) and the tendency to use informal communication mechanisms (Nielsen and Thomsen, 2009) have been identified as reasons for the incipient research conducted so far. At any rate, this state should not be interpreted as a lack of implementation of CSR practices because, as several empirical studies at the international and national level have supported, SMEs carry out more practices than they acknowledge and, thus, they promote. This phenomenon has been referred to in literature as "silent social responsibility" (Jenkins, 2004; Jamali et al., 2009).

Two groups of complementary reasons have been identified to explain the firm's engagement in CSR: on one hand, those of a normative perspective, concerned with the moral correctness of firms and their managers, and on the other, those arising from an instrumental perspective, more related to the traditional performance goals of profitability and business growth. From a strategic approach, most academics agree to prioritize the instrumental focus over the normative perspective; they usually state that those organizations that do not orientate their activities under a CSR philosophy in the short- or medium-term will be at a significant competitive disadvantage. In this regard, for example, different studies have pointed to the adoption of sustainable policies contributing advantages for the company (Porter and Kramer, 2002), either through increasing financial returns (Orlitzky et al., 2003; Miras et al., 2014) or through improving business reputation (Bear et al., 2010; Stanaland et al., 2011).

Aguinis and Glavas (2012) highlighted the gap that exists in the literature in regards to the relationship between CSR and business performance, and encouraged researchers to clarify some of the "mechanisms" that make this association possible. While the relationship between CSR and business performance measures has already been scrutinized in several works, most of them focused on large corporations. Despite it is difficult to suppose that conclusions can be directly extrapolated to the SMEs, at the moment, few studies have been carried out analysing the CSR-firm performance link in SMEs (e.g., Marín and Rubio, 2008; Niehm et al., 2008; Hammann et al., 2009; Sweeney, 2009; Torugsa et al., 2012; Gallardo-Vázquez and Sánchez-Hernández, 2013, 2014a; Turyakira et al., 2014). Of these studies, only a few have tried to measure the CSR practices from a multi-stakeholder perspective (e.g. Hammann et al., 2009; Sweeney, 2009; Torugsa et al., 2012; Turyakira et al., 2014), even though none of them employed a measure of business performance based on the competitiveness that reflected the level of financial and non-financial results achieved in comparison with the most direct competitors. Thereby, none of the reviewed studies considered a performance measure based on the economical–financial position along with the market position regarding quality of products, innovation and customers' satisfaction.

Given this considerations; being aware of the special implementation of CSR practices conducted by reduced dimension businesses; considering the resources limitations of this kind of organizations; and lastly, knowing the owners/managers' confusion about real effects of sustainable behaviours, this investigation aims to clarify the relationship between CSR and competitive performance targeting in the specific context of SMEs (i.e., companies with up to 250 employees). Additionally, delving into the causal relation between the main variables under study and following Surroca et al. (2010) suggestion, this paper investigates whether a firm's capacity to manage stakeholders and achieve competitive advantages has a mediating role between the CSR and the competitive performance of SMEs. Besides, this paper tests the role of firm size as control variable, which impacts on CSR and competitive performance.

To accomplish these tasks, the paper first includes a review of some important theoretical background about the CSR-firm performance link, paying special attention to the SMEs field. The next section also includes the proposition of the hypotheses under analysis. Subsequently, "Methods" section describes the sample, the questionnaire and the statistical process. Finally, after presenting an analysis of results and a discussion of the main findings in "PLS Analysis and Results" and "Discussion" sections, respectively, some conclusions are drawn in the final section. The paper ends warning of some limitations and suggesting future lines of research.

CSR AND FIRM PERFORMANCE LINK

Interest in finding any possible relationship between CSR and business performance[5]emerged more than forty years ago. Most previous studies corroborate that efforts to carry out CSR practices improve firm performance (Beurden and Gössling, 2008). However, the first impression is a field of mixed evidence (Peloza, 2009). Barnett stated in 2007 (p.794) that, "after more than thirty years of research, we cannot clearly conclude whether a one-dollar investment in social initiatives returns more or less than one dollar in benefit to the shareholder." While some studies show a positive relationship (Maron, 2006; Wu, 2006; Rodgers et al., 2013; Gallardo-Vázquez and Sánchez-Hernández, 2014a), others older claim a negative influence (Boyle et al., 1997; Wright and Ferris, 1997); besides, further studies are not able to demonstrate the direction or the sign of the relationship (Barnett and Salomon, 2006).

Margolis and Walsh (2003) also warned of these discrepancies and the consequent confusion when they pointed out the need to understand the conditions under which a corporation's effects impact society before looking for any link between a firm's social and financial performance. Discovering a universal rate of return for CSR activities is virtually impossible; it is necessary to bet on a contingent perspective (Barnett, 2007). Therefore, the model we propose for the specific context of the SMEs not only looks at the direct link between CSR and firm performance, but also aims to clarify how the CSR efforts are rewarded by the different stakeholders.

THE DIRECT RELATIONSHIP BETWEEN CSR AND FIRM PERFORMANCE

Preston and O'Bannon (1997) identify that the theoretical frameworks, in which this relationship is normally underpinned, have two basic differentiating characteristics: the positive, negative or neutral sign of the relationship and the causal sequence of dependent variables. Based on the hypotheses of these authors and on Waddock and Graves's article (1997), seven possible types of CSR-financial performance (FP) relationships could be identified. Of which, most of the empirical outcomes support the social impact hypothesis and lead to reject any negative relationship between CSR and business performance (Beurden and Gössling, 2008).

The underlying theory, which suggests that a higher level of CSR practices leads to a higher level of business performance, is the stakeholder theory (Freeman, 1984). According to the stakeholder theory, the success of an organization depends on the organization's capacity to manage the relationships with its stakeholders. Management of relationships with key business stakeholders has become an essential tool for value generation (Hammann et al., 2009), making the stakeholder theory's interpretation a necessary step in understanding any possible relationship amongst CSR and firm performance (Perrini et al., 2011).

In the field of SMEs, most studies that have analyzed the CSR-performance link have only made suggestions about strategic CSR adoption and implementation, or have explained some theoretical implications in regards to translating the integration of a socially responsible behaviour in business strategy into an improved performance (Moore and Manring, 2009; Tomomi, 2010). Only a few authors have been concerned with providing empirical evidence to corroborate these theoretical implications (Niehm et al., 2008; Hammann et al., 2009; Torugsa et al., 2012; Battaglia et al., 2014; Gallardo-Vázquez and Sánchez-Hernández, 2014a; Turyakira et al., 2014).

From the theoretical perspective of stakeholders, the owners/managers of SMEs are able to more easily express their values towards internal stakeholders and external stakeholders with whom they have a close relationship, than towards more abstract ones like society (Hammann et al., 2009). Despite how SMEs manage their reputation, these organizations are generally better positioned to take advantage of the potential benefits from CSR programmes (Sarbutts, 2003) because they maintain more close, honest and flexible relationships with most of their stakeholders (Russo and Tencati, 2009). The owners/managers able to establish the appropriate objectives and join efforts to accomplish a socially responsible behaviour will contribute to the short-term benefit and to the competitiveness and long-term business growth (Moore and Manring, 2009; Revell et al., 2010).

To achieve these competitive advantages, several authors make their own considerations. Tomomi (2010) noted, for example, that SMEs perceive environmental management as a possible way to provide opportunities for their business and likely cause competitive advantages. Aragón-Correa et al. (2005) highlighted that the commitment to environment and the implementation of measures for its protection can become an important source of competitive advantages for SMEs. On the other hand, Niehm et al. (2008) proved how a company's commitment to the community was directly associated with financial performance.

Under the instrumental approach of stakeholder theory, CSR should be incorporated into business planning by trying to develop business strategies that meet the approval of stakeholders (Parker, 2005). In this way, managers could try to maximize the benefits and value of their businesses while meeting the expectations of their stakeholders (Jensen, 2001). However, this maximization of value cannot be measured only from a financial perspective, but also from a broader approach. Marín and Rubio (2008), for instance, used a multidimensional measure of competitive success, made up of seven dimensions: market share, productivity, solvency, reputation, customer satisfaction, employee satisfaction and competitive position in terms of price, quality and innovation.

From an instrumental approach of stakeholder theory, Hammann et al. (2009) and Sweeney (2009) proposed to examine the relationship between CSR and firm performance considering that socially responsible management enabled companies to gain competitive advantages that add value. In both studies, a similar theoretical model is proposed which basically relates three variables: CSR, measured as the practices of socially responsible management towards different stakeholders; firm performance, as profit improvement; and a last variable which would bring together the potential and expected effects of suitable stakeholder management (improving employee satisfaction, reducing absenteeism, powering image, securing the loyalty of customers and employees, etc.). In both cases, the authors end up stating that CSR implementation improves firm performance through the impact that these practices have on the organization-stakeholder relationships.

In one of the latest studies, Torugsa et al. (2012) also found that the SMEs' capability of managing stakeholders, along with the development of a proactive strategy and the knowledge to achieve a shared vision, are positively associated with a proactive CSR. In turn, their study shows how this proactive CSR causes an improvement in firm financial performance. Gallardo-Vázquez and Sánchez-Hernández (2013), in a survey developed from Spanish micro-enterprises,[6] verified that the positive predisposition of a firm's CEO has a direct impact on competitive success and that innovation positively mediates the relationship between these variables. Another recent study analysing the link between CSR and competitiveness performance has been conducted by Battaglia et al. (2014). Using data from Italian SMEs, which operate in the fashion industry, authors show a strong and positive correlation among several social performance indicators and two competitiveness dimensions: innovation and intangible performance.

Literature reviews and meta-analysis studies about the relationship under discussion have postulated the social impact hypothesis as the most frequent theoretical background in the study of CSR-firm performance link (Orlitzky et al., 2003; Beurden and Gössling, 2008; Aguinis and Glavas, 2012; Miras et al., 2014). In addition, the few studies identified in the context of SMEs leads us to reject any causal relationship that foresees companies will not be positively rewarded by their efforts in CSR. Therefore, since we hold that any relationship that arises between these variables must be made adopting the hypothesis of social impact, that is, expecting positive stakeholders' reactions coming from increasingly socially responsible business behaviour (instrumental approach of stakeholder theory), the first hypothesis for testing in this paper as follows:

H_1. The development of CSR practices enables SMEs to improve their competitive performance.

CSR-FIRM PERFORMANCE: THE MEDIATING ROLE OF RELATIONAL IMPROVEMENTS

From the stakeholder management approach, Jones (1995) stated that it is sensible to expect that CSR contributes to the creation of sustainable competitive advantages that improve business performance and encourage a positive relationship between the variables. Barnett (2007) asserted that the positive impact that CSR could have on business performance primarily would be due to the achievements arisen from an improved socially responsible behaviour and its effect on firm-stakeholder relationships. Marín et al. (2012) claimed that CSR efforts foster the emergence of positive stakeholders' reactions to the company and that this does not only affect a firm's value in general, but it also improves the competitive positioning.

The concern for stakeholders and the communication with them for seeking to ensure a socially responsible behaviour allow firms to develop some particular intangible assets. On the basis of the establishment of relationships and interactions, firms are able to develop resources that will promote the development and maintenance of competitive advantages, which are hard to measure by means of accounting or physical parameters.

Investigation into company-stakeholder reciprocity was already proposed as a future research line by Perrini et al. (2011), who suggested that the usage of mediating variables that represent these interactions would contribute to increase knowledge about the mechanisms through which the CSR practices positively impact on business performance. Different authors have noted that the appearance of empirical discrepancies about the sign identified for the CSR-firm performance link might be due to the omission of certain confounding variables (McWilliams et al., 2006; Orlitzky, 2008). The search for a simple correlation between these variables might not be right because CSR could end up impacting on the income statement through intermediary variables. Thus, as Surroca et al. (2010) tested, the omission of some moderating and mediating variables would justify the lack of evidence to contrast that CSR and performance are significantly associated.

Because of the growing suggestions around the relevance of considering potential intermediary variables (Beurden and Gössling, 2008) and attending to competitive advantages that a firm can get from its interaction with stakeholders, an additional variable has been added into the model: relational improvements.

In agreement with the theoretical approach adopted, there should be positive reciprocity between the firm and stakeholders. Higher CSR efforts should result in higher rewards coming from stakeholders in the form of positive enterprise perceptions. Specifically, as a concept, this construct encompasses the set of business achievements or benefits coming from the manner in which an enterprise manages its relationships with key stakeholders. Through relational improvements variable, the importance of interaction with stakeholders is emphasized, with the relationship being the unit of analysis, which must be understood as the "farming" of strong interactions between two parties that keep any economic or social link and pursue mutual benefits.

Barney (1986) had already found that those organizations with a closer relationship with their stakeholders were able to achieve certain competitive advantages. At the organizational level, the capacity to manage stakeholders has been defined as "the ability to establish trust-based collaborative relationships with a wide variety of stakeholders, especially those with non-economic goals" (Sharma and Vredenburg, 1998, p.735). Since this ability can be considered, from a CSR approach, to improve competitive performance, our model tries to pinpoint the causal effect that the implementation of sustainable management has on performance and how it disrupts the direct CSR-firm performance relationship.

Relational improvements, therefore, could be viewed simultaneously as both an effect and a cause, depending on whether the attention is focused on the impact that CSR practices have on the relationship with stakeholders or on how business performance can be explained based on the positive reaction of stakeholders. To confirm any mediating effect that this variable could play, the following hypothesis has been proposed:

H_2. In SMEs, the relational improvements have a positive mediating effect on the relationship between developed CSR practices and competitive performance.

At last, along with direct and mediating effects proposed, firm size has been considered as a control variable[7]. Several previous studies have contrasted that size has a significant impact on the association between CSR and business performance (Bansal, 2005; Beurden and Gössling, 2008), even when the sample only consists of SMEs (Sweeney, 2009; Torugsa et al., 2012).

METHODS

Sample and data

The random selection of SMEs was done based on companies from Spain listed in the SABI database. The initial sample was 552 companies located in the Murcia region, in southeast Spain. However, during the filtering process, wrongly answered questionnaires, questionnaires completed by companies with fewer than 10 employees (micro-enterprises) or questionnaires from organizations with a special legal form, such as cooperative societies, demanding different treatment were eliminated. Besides, after collecting data, a group of three experts examined all the responses, looking for senseless answers. Finally, the sample comprised 481 SMEs (Table 1).

Table 1: Sample data analysis.

SIZE	Population census		Sample data	
	Number	**%**	**Number**	**%**
Small	4260	87.65	355	73.80
Medium	600	12.35	126	26.20
Total	4860	100.00	481	100.00
	Chi-square			94.624
	Degrees of freedom			1
	Significance level			0.000
	Contingency coefficient			0.1382

Source: DIRCE data (2010).

The data collection process was carried out directly by CSA Consultants Company[8], which was responsible for contacting each of the selected SMEs from December 2010 to February 2011. The information about the variables used for the study was collected through a questionnaire addressed to the companies' managers or the person in charge of the social responsibility issues and it took, on average, approximately 20 minutes to complete. This questionnaire was previously validated through a pilot survey, after which some of the questions were modified and reworded in order to ensure the correct understanding and answer by interviewees, irrespective of the size of their firms[9]. Most questions were measured through a five-point Likert scale, since it has been widely used in surveys conducted about CSR in SMEs (Gallardo-Vázquez and Sánchez-Hernández, 2014b).

To assess non-response bias, Pearson chi-square test and Contingency Coefficient were used to compare the business size of respondent and non-respondent firms, and to evaluate if there was any significant difference. No significant differences were found and it can be concluded, therefore, that, based on size, there is no difference between firms that respond and those that did not.

MEASURES

CSR practices (CSR)

Since the social impact hypothesis arises from the CSR multi-stakeholder perspective, we have considered a CSR measure which gathers items grouped around four key stakeholders (Turker, 2009; Battaglia et al., 2014; Turyakira et al., 2014): environment, employees, society and customers. Specifically, the scale of Lechuga (2012) was chosen as an instrument for measuring the CSR. The author developed and validated a measuring instrument according to the psychometric theory of scales validation. In her study, 57 CSR practices were grouped into six stakeholder categories (employees, customers, suppliers, environment, local community and corporate governance). The author designed and validated a final scale composed of 24 items, grouped into four key stakeholder categories (environment, employees, customers and local community), as specified in Table 2.

Table 2: 'CSR practices in SMEs' scale.

Corporate social responsibility practices	
Indicate your level of agreement with the following statements about environmental practices (1 = strongly disagree, 5 = strongly agree)	
En.Pr.1	Minimize the environmental impact of your business activities
En.Pr.2	Design products and packaging that can be re-used, repaired or recycled
En.Pr.3	Goes voluntarily beyond legal environmental regulations
En.Pr.4	Regularly conducts environmental audits
En.Pr.5	Reuses and recycles materials
En.Pr.6	Adopts measures for ecological design in product/services
En.Pr.7	Implement programmes for the use of alternative energy
En.Pr.8	Implement programmes to reduce water consumption
En.Pr.9	Makes energy-Saving Investments
Indicate your level of agreement with the following statements about practices related to employees (1 = strongly disagree, 5 = strongly agree)	

Em.Pr.1	Employees' interests are taken into account in company decision-making
Em.Pr.2	Support employees who wish to continue or upgrade their education/training
Em.Pr.3	Help the employees find suitable work/life balance (flexible working hours)
Em.Pr.4	Recognizes the importance of stable employment for your employees and society (in the local area)
Em.Pr.5	Develop/Implement regular training programmes
Em.Pr.6	Assess employees work/labour environment on a regular basis
Indicate your level of agreement with the following statements about practices related to local community (1 = strongly disagree, 5 = strongly agree)	
S.Pr.1	Incorporates/includes local community interests in company decision making
S.Pr.2	Support sports or cultural activities in the local community
S.Pr.3	Maintain clear relations with local government authorities
S.Pr.4	The business considers itself to be part of the local community and therefore care about its development/local impact or the improvement of the local infrastructure
S.Pr.5	Support programmes for the disadvantaged
Indicate your level of agreement with the following statements about practices related to customers (1 = strongly disagree, 5 = strongly agree)	
C.Pr.1	Meets its commitments with quality and fair price
C.Pr.2	Inform customers about the proper use of their products and warnings of potential risks
C.Pr.3	Take measures to prevent customer complaints
C.Pr.4	Respond to customer complaints or inquiries

Source: From Lechuga (2012).

In addition to being one of the few alternatives to assess the level of CSR practices in the field of SMEs, it should be noted the structural similarities among this and the scales that other authors have proposed under the stakeholder approach, regardless of organizational size (Turker, 2009; Mishra and Suar, 2010; Turyakira et al., 2014). The validity of each one of the indicators used can be also corroborated according to other CSR measures developed from different approaches, such as the triple bottom line perspective (Torugsa et al., 2012; Gallardo-Vázquez and Sánchez-Hernández, 2014a).

Competitive performance (C.PERF.)

For assessing firm performance, we have opted for an approach similar to the one adopted by Marín et al. (2012) or Gallardo-Vázquez and Sánchez-Hernández (2013), using a variable more focused on assessing competitive performance than on financial performance data. Consequently, the competitiveness has been taken from a two-dimensional perspective, using subjective information about some financial measures on the one hand, and some commercial and technological differentiation items on the other. Thus, after conducting a review of prior studies focused on SMEs in which any measure of competitiveness was used, we designed an initial scale made up of seven items (Table 3). Of these, four were more associated with financial economic performance and three with competitive differentiation.

Table 3: Items of the questionnaire to measure Competitive Performance and Relational Improvements.

Competitive performance	
For the following indicators, compare the results of your company to other competitors (1 = much worse than competitors, 5 = much better than competitors)	
Perf.1	Sales
Perf.2	Market share
Perf.3	Customer satisfaction
Perf.4	Benefits/utility
Perf.5	Profitability
Perf.6	Quality of products
Perf.7	Technological superiority

Relational improvements	
Despite the context of general crisis in the past three years, there have been improvements relating to… (1 = strongly disagree, 5 = strongly agree)	
Imp.1	Service to customers
Imp.2	Relations with customers
Imp.3	Customer loyalty

Imp.4	The firm image
Imp.5	Relations with suppliers
Imp.6	The cost reduction in supplies
Imp.7	The cost reduction in logistic and inventory
Imp.8	The employee satisfaction
Imp.9	Absenteeism
Imp.10	Work environment
Imp.11	The loyalty and morale of employees
Imp.12	The owners and investors satisfaction
Imp.13	Relations with owners
Imp.14	The owners' knowledge about the progress of the firm
Imp.15	Relations with its local community and environment

Source: Competitive performance items also employed by Marín and Rubio (2008), Marín et al. (2012), Gallardo-Vázquez et al. (2013), Gallardo-Vázquez and Sánchez-Hernández (2013, 2014a), among others. Relational Improvements items from Ortiz and Kühne (2008), Iturrioz et al. (2009), Hammann et al. (2009), Battaglia et al. (2014), among others.

Before performing an analysis of the dimensionality of the construct, a global internal consistency test for the seven issues considered as indicators of competitive performance was carried out. Additionally, we contrasted that the elimination of no variable improved the internal consistency of the scale, which reinforces the assumption that these seven indicators were related and they measure an underlying concept. Table 4 shows the results from factor analysis.

Table 4: Factor analysis and refinement of competitive performance scale rotated components ($N = 481$).

Items	**Communality**	**F.1**	**F.2**	**Corr. item-total factor corrected**	**α of subscale if item is eliminated from its factor**
Perf.1	0.653	**0.762**	0.269	0.682	0.818
Perf.2	0.565	**0.706**	0.259	0.609	0.848
Perf.3	0.666	0.267	**0.771**	0.585	0.588

Perf.4	0.802	**0.887**	0.123	0.753	0.787
Perf.5	0.786	**0.880**	0.107	0.735	0.795
Perf.6	0.778	0.116	**0.875**	0.667	0.682
Perf.7	0.607	0.167	**0.761**	0.525	0.757
Variance explained		**50.07%**	**19.33%**		
Total variance explained		**69.40%**			
α of subscale		**0.853**	**0.758**		
α of refined subscale		0.853	0.758		

Bold and shaded values represent the highest loading of each item on the dimensions (factors) identified for the competitive performance scale.

Source: Own elaboration.

The factorial analysis results show how the two dimensions proposed gather indicators which aim to measure a similar concept. One of them collects items about the economic and financial performance compared with competitors, and the other gathers items about the competitive differentiation. Moreover, to corroborate the homogeneity of the questions within each dimension, it was verified that the correlation between each item and its subscale was greater than 0.5 and that the Cronbach's alpha value did not increase after removing it. As can be appreciated in Table 4, no item has been eliminated after the refining process.

Relational improvements (R.IMP.)

For this mediating variable, a multidimensional construct has been designed from a multi-stakeholder organizational management perspective, identifying as first-order dimensions the advantages from managing their relationship with each stakeholder identified for this kind of organization: customers, employees, suppliers, owners and local community. As with the competitive performance construct, we performed a global internal consistency test of the fifteen items (see Table 4) considered initially to measure this construct and selected from a literature review. Table 5 shows the results from factorial analysis of the scale proposed to measure the relational improvements construct.

Table 5: Factor analysis and refinement of relational improvements scale rotated components ($N = 481$).

Items	Communality	F.1	F.2	F.3	F. 4	Corr. item-total factor corrected	α of subscale if item is eliminated from its factor
Imp.1	0.757	**0.825**	0.136	0.217	0.103	0.769	0.823
Imp.2	0.808	**0.842**	0.182	0.217	0.133	0.824	0.810
Imp.3	0.607	**0.727**	0.199	0.170	0.102	0.664	0.851
Imp.4	0.649	**0.776**	0.097	0.180	0.070	0.670	0.847
Imp.5*	0.526	**0.595**	0.322	0.118	0.234	**0.558**	**0.873**
Imp.6	0.898	0.190	0.176	0.106	**0.905**	0.803	–
Imp.7	0.894	0.165	0.138	0.153	**0.908**	0.803	–
Imp.8	0.566	0.410	0.221	**0.574**	0.139	0.587	0.806
Imp.9	0.679	0.114	0.065	**0.792**	0.187	0.558	0.823
Imp.10	0.804	0.382	0.270	**0.765**	0.016	0.781	0.717
Imp.11	0.721	0.185	0.401	**0.723**	0.050	0.685	0.763
Imp.12	0.762	0.234	**0.812**	0.200	0.088	0.739	0.651
Imp.13	0.840	0.185	**0.874**	0.188	0.078	0.820	0.775
Imp.14	0.759	0.099	**0.840**	0.131	0.161	0.740	0.810
Imp.15*	0.500	0.291	**0.573**	0.259	0.139	**0.545**	**0.884**
Variance explained		**44.19%**	**10.75%**	**9.41%**	**7.45%**		
Total variance explained		**71.80%**					
α of subscale		0.869	0.862	**0.825**	**0.890**		
α of refined subscale		**0.873**	**0.884**	0.825	0.890		

Bold and shaded values represent the highest loading of each item on the dimensions (factors) identified for the relational improvements scale.

*Deleted items to refine the scale.

Source: Own elaboration.

From this initial analysis, it can be identified that, in most cases, the dimensions group homogeneous elements, with the exception of items Imp.5 (Improvements relating to firm-suppliers relationships) and Imp.15 (Improvements relating to firm-local community relationships). Although these indicators no have low loadings, it is true that their values are considerably lower than other indicators' loadings. Furthermore, their inclusion in Factor 1 (related to customers) and Factor 2 (related to owners/investors) lacks any theoretical support. Like with the competitive performance, the homogeneity and consistency within each dimension were also checked. As Table 5 shows, the items Imp.5 and Imp.15 have not been measured reliably and disrupt slightly the internal consistency of their respective subscales. This leads us to definitely remove both items and consider that SMEs' managers do not perceive or have difficulties appreciating any improvement related to suppliers and community relationships. In this way, the final scale obtained for relational improvements is composed of 13 variables, grouped in four dimensions, as the result of their relationship with customers, suppliers, employees or owners/investors.

Size

The control variable was measured using, as manifest variables, the natural logarithm of total assets and the natural logarithm of number of employees, both measures taken from 2010. Simultaneously or individually, both measures have often been used in previous empirical studies about CSR and performance (Sweeney, 2009; Blanco et al., 2013).

Methodology

The main methodology used has been the Structural Equation Modelling (SEM). The structural equation analysis has the advantage of analysing the definition of the latent constructs in the context of a group of causal effects. SEM brings together the contributions of econometrics referring to the prediction, along with the psychometric approach, related to the measurement of latent or unobserved variables inferred from indicators or manifest variables (Chin, 1998). There are two different statistical techniques for the SEM: methods based on covariance analysis and methods based on variance analysis.

In this case, we carried out the analysis using a variance-based SEM technique: Partial Least Squares (PLS). We have used this technique, instead of covariance-based models, because it is more suitable for the implementation of predictive studies, which explore complex problems and where previous theoretical background is scarce (Hulland et al., 2010). Moreover, only the PLS technique allows modelling variables of a formative nature (Polites et al., 2012). This technique has been previously used in similar studies (López-Gamero et al., 2008; Blanco et al., 2013; Gallardo-Vázquez et al., 2013; Gallardo-Vázquez and Sánchez-Hernández, 2014b), due to its potential to describe relatively new phenomena from theoretical models and measures without a thorough theoretical background (Chin, 1998).

Before running SEM, the explanatory factorial analysis technique was carried out to second-order variables not validated previously in the literature, with the intention of identifying possible clusters of variables with common meaning. Thus, in addition to choosing contrasted questions from the literature to measure the relational improvements and competitive performance variables, it was necessary to check the reliability of the scales with the data extracted from the sample.

PLS analysis and results

The structural equation analysis has the advantage of analysing the definition of the latent variables in the context of a set of causal relationships. To do this, following the recommendations of some previous authors, such as Barclay et al. (1995), the contrasts should be interpreted distinguishing between the measurement and structural models, applying in each of them the suitable statistical parameters to assess them.

Since the main objective of this paper is the hypothesis contrast between second-order constructs (CSR, R.IMP. and C.PERF.) and a control variable, we are only going to put forward the analysis of the second-order model, having previously checked the validity of the measurement model of first-order according to the two-stage approach (Wright et al., 2012).

On the other hand, according to previous recommendations in the literature, the evaluation of the measurement model was performed as many times as models were considered to contrast each hypothesis. However, as significant differences were not found, we only present the results for the evaluation of the measurement model corresponding to the full model.

Before looking at the results of the estimations performed with the programme PLS-Graph 3.00, an assumption conditioning the analysis of the results should be explained: the CSR and R.IMP. constructs have been considered as second-order formative constructs. In both cases, we understand that the dimensions of these constructs capture heterogeneous aspects and that, therefore, they should be added to develop a measure of performed CSR practices in the first case, and a measure of achieved relational improvements in the second.

Measurement validation

The evaluation of the measurement model presented in Table 6 corresponds to the full model proposed. It is, therefore, a model containing both formative (CSR and R.IMP.) and reflective variables (C.PERF. and SIZE).

Table 6: Measurement validation.

Measurement model				
Indicator		**Weight**	**Loading**	**Residual variance**
CSR construct				
En.Pr.	Practices related to environment	.354***	.666	.557
Em.Pr.	Practices related to employees	.548***	.844	.288
S.Pr.	Practices related to society	.150**	.499	.751
C.Pr.	Practices related to customers	.331***	.684	.532
Competitive performance construct				
E.F.Perf.	*Economic and financial performance*	.495	.788	.380
C.Diff.	*Competitive differentiation performance*	.682	.895	.199
Relational improvements construct				
C.Imp.	Improvements related to customers	.443***	.849	.280
O-I.Imp.	Improvements related to owners and investors	.226***	.697	.513
Em.Imp.	Improvements related to employees	.406***	.858	.264

Su.Imp.	Improvements related to suppliers	.204***	.581	.662
Size construct				
Ln.N.Emp.	Ln of number of employees	.666	.901	.188
Ln.T.Ass.	Ln of total assets	.525	.800	.360

Validation of the formative constructs					
	Indicators	**Original sample estimate**	**Variance Inflation Factor**	**Resampling bootstrap**	
				Mean of subsamples (standard error)	***T*-statistic**
CSR	En.Pr.	.354***	1.214	.355 (.062)	5.740
	Em.Pr.	.548***	1.358	.547 (.063)	8.717
	S.Pr.	.150**	1.199	.146 (.076)	1.980
	C.Pr.	.331***	1.261	.326 (.070)	4.674
R.IMP.	C.Imp.	.443***	1.607	.450 (.081)	5.464
	O-I.Imp.	.226***	1.478	.230 (.072)	3.084
	Em.Imp.	.406***	1.819	.396 (.078)	5.190
	Su.Imp.	.204***	1.222	.198 (.068)	3.012

Validation of the reflective constructs								
	Internal consistency reliability. Composite Reliability (ρ_c)	**Convergent validity (AVE)**	**Discriminant validity AVE$^{1/2}$ > correlations with…**					
			AVE$^{1/2}$	**CSR**	**C.PERF.**	**R.IMP.**	**SIZE**
C.PERF.	.830	.710	.843	.450	1	.477	.218
SIZE	.841	.726	.852	.184	.218	.107	1

*Significant at $p < 0.10$.

**Significant at $p < 0.05$.

***Significant at $p < 0.01$.

Source: Own elaboration.

The most representative parameters, when working with formative character indicators, are the relative weights of each of these on the latent variable (Barclay et al., 1995). Because of this nature, typical of formative constructs, their validity cannot be evaluated using the same criteria as in the reflective models; that is, it cannot be assessed according to whether there is convergent and discriminant validity. Bollen (1989) proposes, instead, to study the formative construct validity by looking at the intensity and statistical significance of the weights of each item, as they give information about the relevance that these measures have on the structural model and on the determination of the latent construct they relate to.

Since, for the case of second-order formative constructs, the weights of the first-order constructs (dimensions) represent causal relationships between these and the former, the weights' statistical significance was analyzed with the same procedure used for evaluating relationships between constructs: the resampling technique of Bootstrap, with which we have generated 5.000 alternative samples from the original data matrix. For each one of these subsamples, PLS re-estimates the parameters and then analyses the consistency of the results and determines if the coefficients obtained provide valid measures of population parameters. The accuracy of these estimates is contrasted using the T test statistic, with $n-1$ degrees of freedom, designed to test the null hypothesis that the estimated parameters from the population are not significantly different from those obtained in the sub-samples.

The results gathered in Table 6 allow us to see that formative constructs incorporated into the models (CSR and R.IMP.) are well-captured by their first-order dimensions: most with weights over 0.2 and significant at the 0.01 level. From an individual analysis of the weights, it is necessary to highlight that, in the case of the CSR construct, the highest weight belongs to practices related to employees. In contrast, the dimension capturing the practices related to society has the lowest weight. With respect to the R.IMP construct, the relational improvements coming from customers and employees have a weight above the others.

Considering the scales "CSR Practices" and "Relational Improvements" as formative variables requires us to analyze the absence of multicollinearity. According to Diamantopoulos and Siguaw (2006), values for Variance Inflation Factor (VIF) over 3.3 suggest the existence of a high multicollinearity.

Furthermore, the data presented in Table 6 confirm the existence of competitive performance and size as reflective constructs consistently measured. In accordance with Carmines and Zeller (1994), the indicators have loadings exceeding 0.707, the composite reliability coefficients (ρ_c) are higher than 0.7 (internal consistency reliability), the average variance extracted values (AVE) exceed 0.5 (convergent validity), and the correlations with any other variable are smaller than the square root of their AVE values (discriminant validity).

Structural validation

The main causal relationship under contrast, where a latent formative variable (CSR) determines a reflective (C.PERF.), represents a specific model known as "redundancy model" (Chin, 1998) (Fig. 1).

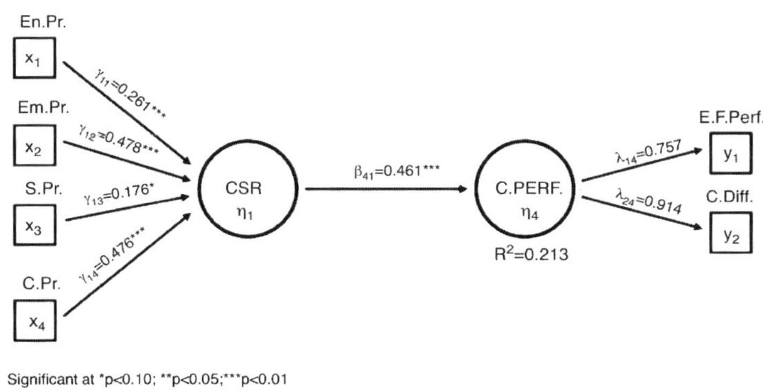

Figure 1: Direct effect of the CSR practices on competitive performance.

Source: Own elaboration.

The assessment of the structural model supports the first hypothesized causal theoretical relationship and, therefore, CSR practices enable SMEs to improve their competitive performance. Attending to the results shown in Table 7, the development of CSR practices has a positive and significant effect on competitive performance (β_{41} = 0.461; significance = 0.01). In addition, it can be argued that the development of CSR practices can predict or explain 21.3% of the competitive performance variance, surpassing the minimum 10% value recommended (Falk and Miller, 1992). The positive value observed for Q^2 indicates that predictive capability of the theoretical model proposed for the contrast of H_1 is adequate.

Table 7: Analysis of the direct effect of the csr practices on competitive performance (Validation of Structural Model).

Validation of Structural Model				
Endogenous latent construct	β_i	**Bootstrap *T*-statistic of the relation**	R^2	**Blindfolding Q^2 measure (standard error)**
C.PERF.	.461***	12.101	.213	0.171 (.087)

*Significant at $p < 0.10$.

**Significant at $p < 0.05$.

***Significant at $p < 0.01$.

Source: Own elaboration.

Having tested the first hypothesis, the isolated mediating effect of the relational improvements has been analyzed (Fig. 2).

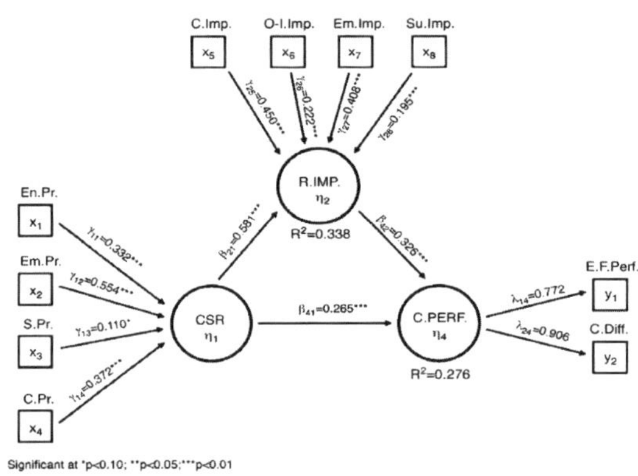

Figure 2: Mediating effect of relational improvements.

Source: Own elaboration.

Again, three indices are examined to evaluate the structural model. First, the relevance measures (Q^2) show that, as before, the predictive capacity is adequate (Table 8). The path coefficients of the relations are higher than the minimum value of 0.2 required, and all are significant at 99%. Despite this, there has been a significant decrease in the coefficient path of the relationship between the constructs C.PERF and CSR, diminishing from 0.461 to 0.265. This decrease was due to the incorporation of the construct R.IMP as a mediating variable. To test this effect, Baron and Kenny (1986) set the following conditions:

1. The partial mediation model should explain more variance of the main dependent variable (C.PERF.) than the model without mediation (R^2before = 0.213, R^2 now = 0.276).
2. There is a significant relationship between the independent variable initially considered (CSR) and the mediating variable (β_{21} = 0.581; significance = 0.01).
3. There is also a significant link between the mediating variable and the variable initially considered as dependent (β_{42} = 0.326; significance = 0.01).
4. Additionally, there must be a significant relationship between the constructs proposed in the direct model (RSE and C.PERF.), although this relationship should have quite slowed down as a result of the inclusion of a mediating variable (β_{41} before = 0.461, β_{41} now = 0.265).

Table 8: Analysis of the mediating effect of relational improvements (Validation of Structural Model).

Validation of Structural Model							
Analysis of…	**Dependent construct**	**Independent construct**	β_i	**Standard Error**	**Bootstrap *T*-statistic of the relation**	R^2	**Blindfolding Q^2 measure (Standard Error)**
The mediating effect of relational improvements	R.IMP.	CSR	.581***	.0297	18.1157	.338	.311 (.082)
	C.PERF.	CSR	.265***	.0526	5.037	.276	.174 (.088)
		R.IMP.	.326***	.0560	5.863		
Size as a control variable	CSR	SIZE	.200***	.0521	3.837	.040	.141 (.094)
	C.PERF.	CSR	.425***	.0416	10.225	.222	.172 (.088)
		SIZE	.134***	.0393	3.406		
The full model proposed	CSR	SIZE	.184***	.0489	3.760	.034	.151 (.098)
	R.IMP.	CSR	.581***	.0321	18.099	.338	.311 (.082)
	C.PERF.	CSR	.235***	.0538	4.365	.291	.174 (.089)
		R.IMP.	.325***	.0560	5.800		
		SIZE	.140***	.0373	3.749		

*Significant at $p < 0.10$.

**Significant at $p < 0.05$.

***Significant at $p < 0.01$.

Source: Own elaboration.

Finally, Chin (1998) suggests a possible measure with which to analyze the mediating effect attending to the variation in the explained variance after the inclusion of a mediating variable in the model:

$$f^2 = \frac{R^2_{\text{included}} - R^2_{\text{excluded}}}{1 - R^2_{\text{included}}}$$

where Rincluded2: variance explained if the mediating variable is introduced; Rexcluded2: variance explained if it is not incorporated in the model.

From data shown in Tables 7 and 8, the f^2 estimated is 0.087, higher than the minimum limit of 0.02, which is considered to state that any variable incorporated into a model contributes to increase the explained variance significantly. Therefore, the second hypothesis is supported and it can be stated that relational improvements have a positive mediating effect on the relationship between developed CSR practices and competitive performance.

Additionally, the role of size as a control variable was checked. As it can be appreciated in Fig. 3, all estimated path coefficients are significant at 99% and there is a greater effect of the control variable on the development of CSR practices than on competitive performance.

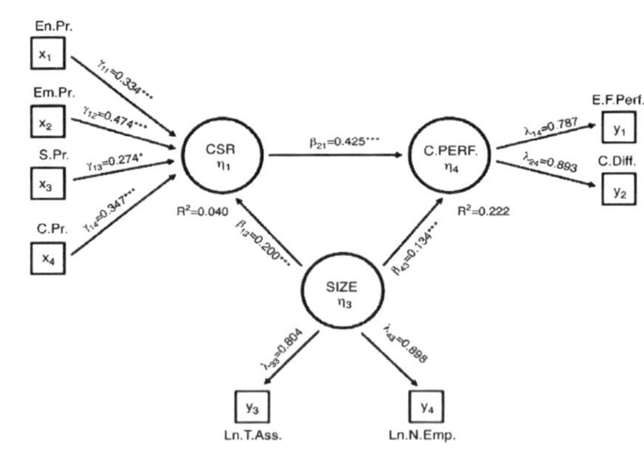

Figure 3: The potential confounding effect of size.

Source: Own elaboration.

According to the results, it can be stated that size, incorporated as an exogenous variable in the model, has a very slight effect on CSR practices and competitive performance. It does not have a decisive incidence on the main relation under study. In any case, through a model assembling firm size along with the mediating effect under study, causal relationships amongst constructs were re-examined in a last analysis (Fig. 4).

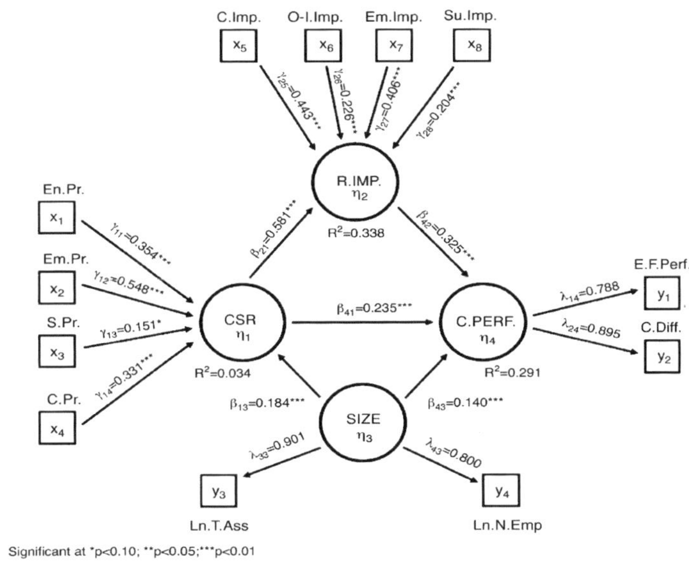

Figure 4: The full model proposed.

Source: Own elaboration.

From the results, first it is necessary to point out that all hypothesized paths are significant at the 0.01 level. The final proposed model explains or contributes to predict 29.1% of the competitive performance variance. Likewise, in agreement with Godos-Díez et al. (2014), two additional and complementary tests were applied to confirm the mediating effect in this full and final model. First, the Sobel *Z* test (1982) was performed to corroborate that this indirect effect was statistically significant ($Z = 5.526$; $p < 0.001$). Secondly, the procedure described by Preacher and Hayes in 2008 to test mediation effects was followed. According to these authors and after applying bootstrapping technique, it can be stated that R.IMP partially mediates the relationship between CSR and C.PERF. and that the indirect impact of CSR on C.PERF has an intensity of 0.189 and is significant, since 0 is not included in its estimated confidence interval of standardized regression (percentile bootstrap at 95%: 0.1327–0.2721).

DISCUSSION

The model finally proposed corroborates that the social impact hypothesis has sufficient empirical support and that the CSR has both a direct and indirect effect on SME competitive performance, precisely as some previous authors had suggested (Aguinis and Glavas, 2012; Miras et al., 2014). Although CSR has been discussed mainly in the context of larger businesses, outcomes indicate that it can be used by SMEs managers as a strategic tool for enhancing the competitiveness (Turyakira et al., 2014).

The evidence found has confirmed the usefulness of the scale developed by Lechuga (2012) to measure the level of CSR practices implemented in SMEs. Along with other kind of alternatives for measuring the socially responsible behaviour of businesses, such as the measures proposed by Marín et al. (2012) or Gallardo-Vázquez et al. (2013), the measure validated in this paper reflects an alternative conceptualization of CSR from the stakeholder view. Thus, our results confirm a scale with an internal structure comparable to the scales of Turker (2009) or Battaglia et al. (2014), with the attractiveness that our measure has been specifically validated for SMEs of any sector.

Moreover, the model estimation and the measurement validation of this formative variable corroborate the relevance that each dimension has to represent a heterogeneous facet of a socially responsible behaviour. Of the four stakeholders that make up this variable (environment, employees, society and customers), the practices attending to employees, environment and customers contribute to the CSR performance to a higher extent and, consequently, practices for these three stakeholders will further enhance the final competitive performance. In this regard, our outcomes confirm those statements that point to employees as the stakeholder with the greatest potential to impact on competitiveness, mainly due to their ability to increase or decrease the levels of productivity of SMEs and their role in the value chain of any company (Jenkins, 2006; Spence, 2007; Hammann et al., 2009; Cegarra-Leiva et al., 2012). On the other hand, in relation to the environment and customers, our results support the need to plan a socially responsible strategy that emphasizes the benefits that these two stakeholders can offer (Murillo and Lozano, 2006). Thus, this work helps to expand the current literature about the positive effect that the commitments with the environment and customers have on business performance (Aragón-Correa et al., 2005; Chavan, 2005; Revell et al., 2010; Gallardo-Vázquez and Sánchez-Hernández, 2014a).

At the same time, a new option to measure competitive performance has been validated. The reliability of its inner model, with similar indicators that have been employed in some previous measures of competitive success or increased competitiveness (Gallardo-Vázquez and Sánchez-Hernández, 2013; Turyakira et al., 2014), allows managers to improve their knowledge about the real effects of CSR practices on firm performance, specifically effects on the competitive performance in the medium- and long-term. According to our expectations, it should be regarded as a reflective construct composed of two main dimensions: one that includes economic and financial aspects, and another that gathers some measures of competitive differentiation. Therefore, our findings are in agreement with the suggestions of Mishra and Suar (2010) about the need to carry out a double evaluation of firm performance, one from a financial point of view and other from a non-financial perspective.

In regards to the inner model of relational improvements, the adequacy of the multi-stakeholder approach used has been confirmed (Porter and Kramer, 2006; Bhattacharya et al., 2009; Mishra and Suar, 2010). Again, our results support the claims of those who advocate that, due to their direct involvement in the value chain, the relationship with employees (internal stakeholder) and customers (external stakeholder) should be carefully managed, because they will have a higher significant impact on corporate performance than other stakeholders (Berman et al., 1999; Hammann et al., 2009). Managers interviewed are unable to appreciate the potential competitive advantages offered by the establishment and maintenance of good and close relationships with the local community, even though these relationships have often been classified as relevant in some investigations carried out in Common Law countries (Lepoutre and Heene, 2006; Niehm et al., 2008; Sweeney, 2009).

Related to causal effects, the development of CSR practices has a positive influence on competitive performance both directly and indirectly.

Therefore, according to Zhao et al. (2010), there is a *complementary mediation* (i.e. mediated effect and direct effect both exist and point in the same direction). This partial mediation clarifies how the satisfaction of stakeholders' expectations allows for an increase in the potential benefits that SMEs can obtain, thanks to their good relationships with internal and external stakeholders (Perrini, 2006; Worthington et al., 2006; Fisher et al., 2009; Perrini et al., 2011). Certain previous researchers have pointed out that social responsibility policies contribute to achieve competitive advantages (Porter and Kramer, 2006; Moneva et al., 2007). This work verifies Porter and Kramer's suggestions (2006), since it shows how the integration of CSR within SMEs strategy benefits stakeholders and how, in turn, these internal and external agents react in favour of each company, offering new competitive opportunities. In this regard, it is necessary to understand the true value that CSR generates for the company, by way of shared value. This value emerges to redefine and innovate products or markets; it is a value that provides improvements in the value chain, a value that, ultimately, is always generated considering the interaction between the company and the stakeholders (Porter and Kramer, 2011).

Finally, sometimes it has been argued that one of the main advantages that smaller entities could have is attributed to the inverse relationship between size and proximity to stakeholders (Russo and Tencati, 2009; Hammann et al., 2009; Torugsa et al., 2012). Thanks to their better positioning to establish and maintain close links with their key stakeholders, smaller firms would have greater opportunities to take advantage of some of the competitive benefits from the CSR (Niehm et al., 2008; Iturrioz et al., 2009; Battaglia et al., 2014). In short, this research shows that, regardless of their size, those SMEs that develop socially responsible practices favour positive stakeholder reactions and improve its competitiveness (Lepoutre and Heene, 2006).

CONCLUSION, LIMITATIONS, AND FUTURE RESEARCH

One of the main trends of research on CSR has been its potential association with the financial performance. Most authors suggest the social impact hypothesis as a starting point to analyze this association, however the evidence is still unclear, with only a few studies focused in the SMEs field. For this reason, this paper aimed to verify whether, in the field of SMEs, the development of CSR practices also influences business performance. To tackle this objective, empirical data were drawn primarily from personal interviews carried out in early 2011 with more than 481 Spanish SMEs managers. Thus, the large sample of SME's enables us to argue our findings from a broad perspective and close to business reality.

Firstly, the research conducted offers a new way to evaluate the performance of socially responsible companies, adapted to SMEs. Along with the scale proposed by Gallardo-Vázquez et al. (2013), this paper provides to Spanish managers an alternative tool for starting to learn how to embrace and evaluate the CSR: a scale. According to Ortiz and Kühne (2008), the strategic decisions surrounding the socially responsible behaviour of a company can become a source of competitive advantage; thus, managers and owners have the opportunity to take advantage of that which the development of CSR practices provides. On the basis of stakeholder theory, this scale can be considered as a guide or easy-to-use checklist, which involves 24 socially responsible practices with four key stakeholders (in order of relative relevance): employees, environment, customers and society.

According to outcomes reached, the priority management of employee and customer issues contributes more to the achievement of competitive performance than the management of environmental practices and, especially, the development of practices related to society. As Hammann et al. (2009) claimed, the priority management that SMEs use with their stakeholders will be crucial to value generation. The role of relational improvements as a mediating variable emphasizes the relevance of the good management of employees to improve business performance and highlights the importance of this stakeholder in the CSR of SMEs (Turker, 2009). On the contrary, regardless of the presence of the mediating variable, the CSR practices related to society are those that contribute least to increase competitive performance.

The contrast of the full model has allowed us to provide evidence to support the hypothesis of social impact in the context of SMEs. Considering the mediating effect or not, results confirm the existence of a positive and significant causal relationship between the level of CSR practices and competitive performance. On this point, it is necessary to highlight that the inclusion of a mediating variable in the model weakens the relationship intensity between CSR and performance, enough to state that there is a partial mediation. In this regard, as Surroca et al. noted (2010), the debate around CSR and business performance linkage is open. This paper exemplifies that there are some omitted variables that should be considered in models proposed to explain the relationship between these constructs.

To sum up, it has been confirmed that the development of CSR practices promotes strengthening of linkages that SMEs have with their stakeholders and that this improvement, in turn, positively affects competitiveness (Fitjar, 2011; Battaglia et al., 2014; Turyakira et al., 2014). At the same time, the validity of the CSR variable used and the relevance of taking into account the effect of the relational improvements confirm the suitability of the stakeholder theory as the main approach to analyze the relationship between CSR and competitiveness (Perrini et al., 2011).

The results and conclusions drawn from this study should be assumed considering certain limitations. First, the evidence found can only be extrapolated to the SME field, mainly to small enterprises; comparisons to research focused on larger firms should be made with caution. Furthermore, because the strategy of this type of organization is highly conditioned by the attitude and values of the owners/managers, results may differ significantly for a sample of Anglo-Saxon companies. Another limitation is related to respondents. Although most collected information for the quantitative study was obtained from interviews with those responsible for socially responsible management, it should be noted that the data method relied on respondents' subjective reporting of their firm's CSR practices, relational improvements and competitive performance.

Other most common limitations are those deriving from the measurement of the constructs or from the methodology used. For instance, while in this paper the theoretical approach of stakeholders has been used to measure the level of CSR practices, other studies contemplating SMEs have opted for a more classical approach of CSR, focusing on the triple bottom line (Torugsa et al., 2012; Gallardo-Vázquez et al., 2013). Related to structural equation modelling applied and our data, one of the main constraints of our analysis could be the potential endogeneity problem between CSR and competitive performance (García-Castro et al., 2010). However, we understand that the evaluation of the level of CSR performance is not more than the result of socially responsible policies and strategies necessarily undertaken in advance and that, on the other hand, our measure of competitive performance reflects a momentary assessment of business performance. In any case, we recommend the use of lagged performance indicators to solve this limitation.

Finally, while in this paper only the mediating effect of the SME relational capacity has been considered, studies carried out in large companies suggest that there could be other intangible mediating factors, such as human capital or reputation (Surroca et al., 2010), innovation (Blanco et al., 2013) or advertising intensity (Hull and Rothenberg, 2008; Luo and Bhattacharya, 2009). Moreover, we are aware some researchers have considered different organizative characteristics in addition to firm size as control variables, such as the sector, the level of profitability, the leverage or the family character, among others. For instance, it would be expected that enterprises belonging in sectors with a high social or environmental risk had stronger causal effects and that, therefore, a stratification of the sample by sectors helped to assess in which sectors the CSR has more relevance as a source of competitive advantages (Gallardo-Vázquez and Sánchez-Hernández, 2013). In this regard, as a moderator variable, it could be studied if companies belonging to more CSR sensitive sectors are able to achieve a higher firm performance or, conversely, if CSR practices do not offer competitive advantages in this kind of sectors, considering that any sustainable behaviour would become normalized and it would not offer new differentiation opportunities.

In addition, to overcome some of limitations above, future research could be carried out selecting a sample of non-Latin SMEs. In this way, the legal origin could be evaluated as a control variable. It would also be advisable to combine subjective measures of competitive performance with some kind of objective measure of financial performance. This would let us test whether the causal relationship found from respondents' perceptions is consistent with the real economic and financial situation of their companies. To finish, focusing on the structural model, the inclusion of certain mediating variables identified in the literature could help improve the total variance explained of business performance and design a more detailed strategic plan.

REFERENCES

1. Aguinis, H., Glavas, A., 2012. What we know and don't know about corporate social responsibility: a review and research agenda. J. Manage. 38 (4), 932---968.
2. Aragón-Correa, J.A., García-Morales, V.J., Hurtado-Torres, N.E., 2005. Un Modelo Explicativo de las Estrategias Medioambientales Avanzadas para Pequena˜ s y Medianas Empresas y su Influencia en los Resultados. Cuadernos Econ. Dirección Empresa 25, 29---52.
3. Bansal, P., 2005. Evolving sustainably: a longitudinal study of corporate sustainable development. Strategic Manage. J. 26 (3), 197---218.
4. Barclay, D., Higgins, C., Thompson, R., 1995. The partial least squares (PLS) approach to causal modeling: personal computer adoption and use as an illustration. Technol. Stud. 2 (2), 285---309 (Special Issue on Research Methodology).
5. Barnett, M.L., Salomon, R.M., 2006. Beyond dichotomy: the curvilinear relationship between social responsibility and financial performance. Strateg. Manage. J. 27 (11), 1101.
6. Barnett, M.L., 2007. Stakeholder influence capacity and the variability of financial returns to corporate social responsibility. Acad. Manage. Rev. 32 (3), 794---816.
7. Barney, J., 1986. Organizational culture; can it be a source sustained competitive advantage? Acad. Manage. Rev. 11 (3), 656---665.
8. Baron, R.M., Kenny, D.A., 1986. The moderator-mediator variable distinction in social psychological research: conceptual, strategic, and statistical considerations. J. Pers. Soc. Psychol. 51 (6), 1173---1182.
9. Battaglia, M., Testa, F., Bianchi, L., Iraldo, F., Frey, M., 2014. Corporate social responsibility and competitiveness within SMEs of the fashion industry: evidence from Italy and France. Sustainability 6 (2), 872---893.
10. Bear, S., Rahman, N., Post, C., 2010. The impact of board diversity

and gender composition on corporate social responsibility and firm reputation. J. Bus. Ethics 97 (2), 207---221.

11. Berman, S.L., Wicks, S.C., Kotha, S., Jones, T.M., 1999. Does stakeholder orientation matter? The relationship between stakeholder management models and firm performance. Acad.
12. Manage. J. 42 (5), 488---509.
13. Beurden, P., Gössling, T., 2008. The worth of values --- a literature review on the relation between corporate social and financial performance. J. Bus. Ethics 82 (2), 407---424.
14. Bhattacharya, B.B., Sen, S., Korschun, D., 2009. Strengthening stakeholder---company relationships through mutually beneficial corporate social responsibility initiatives. J. Bus. Ethics 85 (2 Suppl.), 257---272.
15. Blanco, B., Guillamon-Saorin, E., Guiral, A., 2013. Do non socially responsible companies achieve legitimacy through socially responsible actions? The mediating effect of innovation. J. Bus. Ethics 117 (1), 67---83.
16. Bollen, K.A., 1989. Structural Equations with Latent Variables. Wiley, New York.
17. Boyle, E.J., Higgins, M.M., Rhee, S.G., 1997. Stock market reaction to ethical initiatives of defense contractors: theory and evidence. Crit. Perspect. Account. 8, 541---561.
18. Carmines, E.G., Zeller, R.H., 1994. Reliability and validity assessment. In: Lewis-Beck, M.S. (Ed.), Basic Measurement. Sage Publication, London, pp. 1---58.
19. Cegarra-Leiva, D., Sánchez-Vidal, E., Cegarra-Navarro, J.G., 2012. Understanding the link between work life balance practices and organisational outcomes in SMEs. The mediating effect of a supportive culture. Pers. Rev. 41 (3), 359---379.
20. Chavan, M., 2005. An appraisal of environment management systems: a competitive advantage for small businesses. Manag. Environ. Qual. 16 (5), 444---463.
21. Chin, W.W., 1998. The partial least squares approach to structural equation modeling. In: Marcoulides, G.A. (Ed.), Modern Methods for Business Research. Lawrence Erlbaum Associates, Mahwah,

New Jersey, pp. 295---336.

22. Diamantopoulos, A., Siguaw, J., 2006. Formative versus reflective indicators in organizational measure development: a comparison and empirical illustration. Br. J. Manage. 17 (4), 263---282.
23. Falk, R.F., Miller, N.B., 1992. A Primer for Soft Modelling. The University of Akron, Akron, OH.
24. Fisher, K., Geenen, J., Jurcevic, M., McClintock, K., Davis, G., 2009.
25. Applying asset-based community development as a strategy for CSR: a Canadian perspective on a win---win for stakeholders and SMEs. Bus. Ethics: Eur. Rev. 18 (1), 66---82.
26. Fitjar, R.D., 2011. Little big firms? Corporate social responsibility in small businesses that do not compete against big ones. Bus. Ethics: Eur. Rev. 20 (1), 30---44.
27. Freeman, R.E., 1984. Strategic Management: a Stakeholder Approach. Pitman, Boston.
28. Gallardo-Vázquez, D., Sánchez-Hernández, M.I., 2013. Análisis de la Incidencia de la Responsabilidad Social Empresarial en el Éxito Competitivo de las Microempresas y el Papel de la Innovación. Univ. Bus. Rev. 38 (segundo trimestre), 14---31.
29. Gallardo-Vázquez, D., Sánchez-Hernández, M.I., Corchuelo Martínez-Arzúa, M.B., 2013. Validación de un Instrumento de Medida para la Relación entre la Orientación a la Responsabilidad Social Corporativa y otras Variables Estratégicas de la Empresa. Span. Account. Rev. 16 (1), 11---23.
30. Gallardo-Vázquez, D., Sánchez-Hernández, M.I., 2014a. Measuring corporate social responsibility for competitive success at a regional level. J. Cleaner Prod. 72 (1), 14---22.
31. Gallardo-Vázquez, D., Sánchez-Hernández, M.I., 2014b. Structural analysis of the strategic orientation to environmental protection in SMEs. Bus. Res. Q. 17 (2), 115---128.
32. García-Castro, R., Arino,˜ M., Canela, M., 2010. Does social performance really lead to financial performance? Accounting for endogeneity. J. Bus. Ethics 92 (1), 107---126.
33. Godos-Díez, J.L., Fernández-Gago, R., Cabeza-García, L.,

MartínezCampillo, A., 2014. Determinants of CSR practices: analysis of the influence of ownership and the management profile mediating effect. Span. J. Finance Account. 43 (1), 47---90.

34. Hammann, E., Habisch, A., Pechlaner, H., 2009. Values that create value: socially responsible business practices in SMEs --- empirical evidence from German companies. Bus. Ethics: Eur. Rev. 18 (1), 37---51.
35. Hull, C.E., Rothenberg, S., 2008. Firm performance: the interactions of corporate social performance with innovation and industry differentiation. Strateg. Manage. J. 29 (7), 781---789.
36. Hulland, J., Ryan, M.J., Rayner, R.K., 2010. Modeling customer satisfaction: a comparative performance evaluation of covariance structure analysis versus partial least squares. In: Esposito Vinzi, V., Chin, W.W., Henseler, J., Wang, H. (Eds.), Handbook of Partial Least Squares: Concepts, Methods and Applications. Springer Handbooks, Berlin, pp. 307---325.
37. Iturrioz, C., Aragón, C., Narbaiza, L., Ibanez,˜ A., 2009. Social responsibility in SMEs: a source of business value. Soc. Responsibility J. 5 (3), 423---434.
38. Jamali, D., Zanhour, M., Keshishian, T., 2009. Peculiar strengths and relational attributes of SMEs in the context of CSR. J. Bus. Ethics 87 (3), 355---377.
39. Jenkins, H., 2004. A critique of conventional CSR theory: an SME perspective. J. Gen. Manage. 29 (4), 37---57.
40. Jenkins, H., 2006. Small business champions for corporate social responsibility. J. Bus. Ethics 67 (3), 241---256.
41. Jensen, M., 2001. Value maximization, stakeholder theory and the corporate objective function. J. Appl. Corp. Finance 14 (3),
42. 8---21.
43. Jones, T.M., 1995. Instrumental stakeholder theory: a synthesis of ethics and economics. Acad. Manage. Rev. 20 (2), 404---437.
44. Lechuga, P., (PhD thesis) 2012. Design and Validation of an Instrument for Measuring the Social Responsibility Practices in Small and Medium Enterprises (written in Spanish). University of Cadiz.

45. Lepoutre, J., Heene, A., 2006. Investigating the impact of firm size on small business social responsibility: a critical review. J. Bus. Ethics 67 (3), 257---273.
46. López-Gamero, M.D., Molina-Azorín, J.F., Claver-Cortés, E., 2008. Análisis de los factores que condicionan la percepción del directivo sobre el medio ambiente. Un estudio Qual/Quan. Cuadernos Econ. Dirección Empresa 37, 123---172.
47. Luo, X., Bhattacharya, C.B., 2009. The debate over doing good: corporate social performance, strategic marketing levers, and firm-idiosyncratic risk. J. Mark. 73 (6), 198---213.
48. Margolis, J.D., Walsh, J.P., 2003. Misery loves companies: rethinking social initiatives by business. Adm. Sci. Q. 48 (2), 263---305.
49. Marín, L., Rubio, A., 2008. ¿Moda o Factor Competitivo? Un Estudio Empírico de Responsabilidad Social Corporativa en PYMEs. Inf. Comercial Espanol˜ a Rev. Economía 842, 177---193.
50. Marín, L., Rubio, A., Ruiz de Maya, S., 2012. Competitiveness as a strategic outcome of corporate social responsibility. Corp. Soc. Responsibility Environ. Manage. 19 (6), 364---376.
51. Mark-Herbert, C., Von Schantz, C., 2007. Communicating corporate social responsibility. Electron. J. Bus. Organ. 12 (2), 4---11.
52. Maron, I.Y., 2006. Toward a unified theory of the CSP-CFP link. J. Bus. Ethics 67 (2), 191---200.
53. McWilliams, A., Siegal, D., Wright, P., 2006. Guest Editors' Introduction: "corporate social responsibility: strategic implications. J. Manage. Stud. 43 (1), 1---18.
54. Miras, M.M., Carrasco, A., Escobar, B., 2014. Corporate social responsibility and financial performance: a meta-analysis. Span. J. Finance Account. 43 (2), 193---215.
55. Mishra, S., Suar, D., 2010. Does corporate social responsibility influence firm performance of Indian companies? J. Bus. Ethics 95 (4), 571---601.
56. Moneva, J.M., Rivera, J.M., Munoz,˜ M.J., 2007. The corporate stakeholder commitment and social and financial performance. Ind. Manage. Data Syst. 107 (1), 84---102.
57. Moore, S.B., Manring, S.L., 2009. Strategy development in small

and medium sized enterprises for sustainability and increased value creation. J. Cleaner Prod. 17 (2), 276---282.

58. Murillo, D., Lozano, J., 2006. SMEs and CSR: An approach to CSR in their own words. J. Bus. Ethics 67 (3), 227---240.
59. Niehm, L.S., Swinney, J., Miller, N.J., 2008. Community social responsibility and its consequences for family business performance. J. Small Bus. Manage. 46 (3), 331---350.
60. Nielsen, A.E., Thomsen, C., 2009. Investigating CSR communication in SMEs: a case study among danish middle managers. Bus. Ethics: Eur. Rev. 18 (1), 83---93.
61. Orlitzky, M., Schmidt, F.L., Rynes, S.L., 2003. Corporate social and financial performance: a meta-analysis. Organ. Stud. 24 (3), 403---441.
62. Orlitzky, M., 2008. Corporate social performance and financial performance: a research synthesis. In: Crane, A., McWilliams, A., Matten, D., Moon, J., Siegel, D.S. (Eds.), The Oxford Handbook of CSR. University Press, Oxford, pp. 113---134.
63. Orlitzky, M., Siegel, D.S., Waldman, D.A., 2011. Strategic corporate social responsibility and environmental sustainability. Bus. Soc. 50 (1), 6---27.
64. Ortiz, D., Kühne, S., 2008. Implementing responsible business behavior from a strategic management perspective: developing a framework for Austrian SMEs. J. Bus. Ethics 82 (2), 463---475.
65. Parker, L., 2005. Social and environmental accountability research: a view from the commentary box. Account. Auditing Account. J. 18 (6), 842---849.
66. Peloza, J., 2009. The challenge of measuring financial impacts from investments in corporate social performance. J. Manage. 35 (6), 1518---1541.
67. Perrini, F., 2006. SMEs and CSR theory: evidence and implications from an Italian perspective. J. Bus. Ethics 67 (3), 305---316.
68. Perrini, F., Russo, A., Tencati, A., Vurro, C., 2011. Deconstructing the relationship between corporate social and financial performance. J. Bus. Ethics 102 (1 Supplement), 59---76.
69. Polites, G.L., Roberts, N., Thatcher, J., 2012. Conceptualizing

models using multidimensional constructs: a review and guidelines for their use. Eur. J. Inf. Syst. 21, 22---48.

70. Porter, M., Kramer, M., 2002. The competitive advantage of corporate philanthropy. Harv. Bus. Rev. 80 (12), 56---68.
71. Porter, M.E., Kramer, M.R., 2006. Strategy and society: the link between competitive advantage and corporate social responsibility. Harv. Bus. Rev. 84 (12), 78---92.
72. Porter, M.E., Kramer, M.R., 2011. Creating shared value. Harv. Bus. Rev. 89 (1/2), 62---77.
73. Preacher, K.J., Hayes, A.F., 2008. Asymptotic and resampling strategies for assessing and comparing indirect effects in multiple mediator models. Behav. Res. Methods 40 (3), 879---891.
74. Preston, L.E., O'Bannon, D.P., 1997. The corporate social-financial performance relationship. A typology and analysis. Bus. Soc. 36 (4), 419---429.
75. Revell, A., Stokes, D., Chen, H., 2010. Small businesses and the environment: turning over a new leaf? Bus. Strategy Environ. 19 (5), 273---288.
76. Rodgers, W., Choy, H.L., Guiral, A., 2013. Do investors value a firm's commitment to social activities? J. Bus. Ethics 114 (4), 607---623.
77. Russo, A., Tencati, A., 2009. Formal vs. informal CSR strategies: evidence from italian micro, small, medium-sized, and large firms. J. Bus. Ethics 85 (2 Supplement), 339---353.
78. Sarbutts, N., 2003. Can SMEs 'do' CSR? A practitioner's view of the ways small-and medium-sized enterprises are able to manage reputation through corporate social responsibility. J. Commun. Manage. 7 (4), 340---347.
79. Sharma, S., Vredenburg, H., 1998. Proactive corporate environmental strategy and the development of competitively valuable organizational capabilities. Strateg. Manage. J. 19 (8), 729---753.
80. Sobel, M.E., 1982. Asymptotic confidence intervals for indirect effects in structural equation models. In: Leinhardt, S. (Ed.), Sociological Methodology. American Sociological Association,

Washington DC, pp. 290---312.

81. Spence, L.J., 2007. CSR and small business in a European policy context: The five 'C's of CSR and small business research agenda 2007. Bus. Soc. Rev. 112 (4), 533---552.
82. Stanaland, A.J.S., Lwin, M.O., Murphy, P.E., 2011. Consumer perceptions of the antecedents and consequences of corporate social responsibility. J. Bus. Ethics 102 (1), 47---55.
83. Surroca, J., Tribo, J.A., Waddock, S., 2010. Corporate responsibility and financial performance: the role of intangible resources. Strateg. Manage. J. 31 (5), 463---490.
84. Sweeney, L., (PhD thesis) 2009. A Study of Current Practice of Corporate Social Responsibility (CSR) and an Examination of the Relationship between CSR and Financial Performance Using Structural Equation Modelling (SEM). Dublin Institute of Technology.
85. Tomomi, T., 2010. Environmental management strategy for small and medium-sized enterprises: why do SMBs practice environmental management? Asian Bus. Manage. 9 (2),
86. 265---280.
87. Torugsa, N.A., O'Donohue, W., Hecker, R., 2012. Capabilities, proactive CSR and financial performance in SMEs: empirical evidence from an Australian manufacturing industry sector. Jo. Bus. Ethics 109 (4), 483---500.
88. Turker, D., 2009. Measuring corporate social responsibility: a scale development study. J. Bus. Ethics 85 (4), 411---427.
89. Turyakira, P., Venter, E., Smith, E., 2014. The impact of corporate social responsibility factors on the competitiveness of small and medium-sized enterprises. S. Afr. J. Econ. Manage. Sci. 17 (1), 157---172.
90. Waddock, S., Graves, S.B., 1997. The corporate social performancefinancial performance link. Strateg. Manage. J. 18 (4), 303---319.
91. Worthington, I., Ram, M., Jones, T., 2006. 'Giving something back': a study of corporate social responsibility in UK South Asian small enterprises. Bus. Ethics 15 (1), 95---108.

92. Wright, R.T., Campbell, D.E., Thatcher, J.B., Roberts, N., 2012. Operationalizing multidimensional constructs in structural equation modeling: recommendations for IS research. Commun.
93. Assoc. Inf. Syst. 30, article 23.
94. Wright, P., Ferris, S.P., 1997. Agency conflict and corporate strategy: the effect of divestment on corporate value. Strategy Manage. J. 18 (1), 77---83.
95. Wu, M.L., 2006. Corporate social performance, corporate financial performance, and firm size: a meta-analysis. J. Am. Acad. Bus. 8 (1), 163---171.
96. Zhao, X., Lynch, J.G., Quimei, C., 2010. Reconsidering Baron and Kenny: myths and truths about mediation analysis. J. Consum. Res. 37 (2), 197---206.

REFERENCES

1. F. Ayres, Theory and Problems of Differential Equations, Schaum's, McGraw-Hill, 1999
2. Barna, A, A. Whinston and F. Yin. Value and Productivity in the Internet Economy. Computer, May, 2000.
3. Barney, D. E-Comm Intelligence Report. Network World, February 28, 2000.
4. Barros, O. Operations Research. Vol 2: Models. Editorial Universitaria, Chile, 1982.
5. Barros, O. Business Process Reengineering, Dolmen, 2nd edition, 1995
6. Barros, O. Information Technology and its use in Management, McGraw-Hill, 1998
7. Barros, O. Business Process Redesign Using Patterns. Dolmen Editorial, 2000.

8. M.Braun, Diferencialesy Equations Applications, Grupo Editorial Iberoamerica, Mexico, 1990
9. Brynjofsson, Erik and Loring Hitt. Paradox Lost? Firm-Level Evidence on the Returns to Information Systems Spending. Management Science, Vol 42, No. 4, April, 1996.
10. Computerworld. Staples.com Customer Service Makes No. 1, September 4, 2000.
11. CTO FirstMover. First Movers Exchange. May 15, 2000.
12. Champy, J. Trends That Never Died: TQM & Reengineering. Computerworld, May 20, 2000.
13. Evans, Ph. And TS Wurster. Getting Real About Virtual Commerce. Harvard Business Review, November-December 1999.
14. Furger, R. Building on the Bazaar website. PC World Chile, May, 2000.
15. F.Florent, et systèmes Equations differentiels, Vuibert, Paris, 1978
16. Gibney Jr, F. Enron Plays the Pipers. Time, August 28, 2000.
17. RJ Gordon Does the "New Economy" Measure up to the Great Inventions of the Past? NBER Working Book No. W7833, August, 2000.
18. Grygo, E. Exchange Future Mixed. Infoworld, June 5, 2000.
19. Gutiérrez, N. Process Design Segmenting the customer base of a commercial bank using data mining techniques. Memory Industrial Engineer, Department. Industry Engineering, University of Chile, 2001.
20. Han, J. and M. Kamber. DataMining: Concepts & Techniques. Morgan Kaufmann Publishers, 2001.
21. Hamel, G. and CK Prahalad. Competing for the Future. Harvard Business School Press, 1994.
22. Hiebeler, R. Kelly and TB Ch. Ketterman. Best Practices. Simon & Schuster, 1998.
23. Infoworld. Brick-and-Mortars E-commerce Take the Plunge. April 3, 2000.

24. Jensen, MC and WH Meckling. Theory of the Firm: Management Behavior, Agency Costs and Ownership Structure. Journal of Financial Economics 3, 1976, pp. 305- 360.
25. Jensen, MC. Organization Theory and Methodology. LVII The Accounting Review, 1983, pp. 319-339.
26. Katz, ML and C. Shapiro. Network Externalities, Competition, and Compatibility. The American Economic Review, 75, No. 3, pp. 425-441, 1985.
27. Keen, PGW Back to Processes. Computerworld, May 8, 2000.
28. J. Kuntzmann, Systèmes differentiels, Hermann, Paris, 1967
29. Klemperer, P. Competition When Consumers Have Switching Costs. An Overview with Applications to Industrial Organizations, Macroeconomics, and International Trade. Review of Economic Studies 62, pp.515-539, 1995.
30. Los Angeles Times. Web Retailers Lead in Internet Job Cuts. August 25, 2000.
31. Mintzberg, H. Structure in 5's: A Synthesis of the Research on Organization Design. Management Science 26, 3, 1980, pp. 322-341.
32. Rohlfs, J. A theory of Interdependent Demand for a Communication Services. Bell Journal of Economics 5, No. 1, pp. 16-37, 1974.
33. Rosencranse, L. Andersen, Sun and iPlanet Set up Venture Procurement. Computerworld, May 1, 2000.
34. Schultz, B. E-Comm End to End. Network World, February 28, 2000.
35. Schwartz, E., D. and E. Neel Grygo. New World Economic Order. InfoWorld, July 17, 2000.
36. Sherman, E. Web Delivery in an Hour. Computerworld, March 6, 2000.
37. Simmons, G.F., Differential Equations, Mc Graw Hill, 1979
38. Syken, B. 25 Best E-commerce Sites. Time, August 21, 2000.
39. Taylor, C. All Boxed in. Time, September 4, 2000.
40. Taylor, C. Till to Drop Bot. Time, August 21, 2000
41. Werbach, K. Syndication: The Emerging Model for Business in the Internet Era. Harvard Business Review, May-June 2000.

42. Williamson, OE Markets and Hierarchies. Tree Press, NY, 1981.
43. Yager, T. Microsoft.Net Impact. InfoWorld, September 4, pp.45-47, 2001.

CITATIONS

CHAPTER 7

Ali Gohary. Kambiz Heidarzadeh Hanzaee. Personality Traits as Predictors of Shopping Motivations and Behaviors: A Canonical Correlation Analysis, https://doi.org/10.1016/j.aebj.2014.10.001

CHAPTER 8

Syeda Anam Hassan, Khalid Zaman, Shah Gul, The Relationship between Growth-Inequality-Poverty Triangle and Environmental Degradation: Unveiling the Reality, Arab Economic and Business Journal, Volume 10, Issue 1, 2015, Pages 57-71, ISSN 2214-4625, http://dx.doi.org/10.1016/j.aebj.2014.05.007.

CHAPTER 9

Isabel Buil, Sara Catalán, Eva Martínez, The importance of corporate brand identity in business management: An application to the UK banking sector, BRQ Business Research Quarterly, Volume 19, Issue 1, 2016, Pages 3-12, ISSN 2340-9436, http://dx.doi.org/10.1016/j.brq.2014.11.001.

CHAPTER 10

Luis Ángel Guerras-Martín, Anoop Madhok, Ángeles Montoro-Sánchez, The evolution of strategic management research: Recent trends and current directions, BRQ Business Research Quarterly, Volume 17, Issue 2, 2014, Pages 69-76, ISSN 2340-9436, http://dx.doi.org/10.1016/j.brq.2014.03.001.

CHAPTER 11

Elisabet Garrido, Jaime Gomez, Juan P. Maicas, Raquel Orcos, The institution-based view of strategy: How to measure it, BRQ Business Research Quarterly, Volume 17, Issue 2, 2014, Pages 82-101, ISSN 2340-9436, http://dx.doi.org/10.1016/j.brq.2013.11.001.

CHAPTER 12

Ester Gras-Gil, Mercedes Palacios Manzano, Joaquín Hernández Fernández, Investigating the relationship between corporate social responsibility and earnings management: Evidence from Spain, BRQ Business Research Quarterly, Volume 19, Issue 4, 2016, Pages 289-299, ISSN 2340-9436, http://dx.doi.org/10.1016/j.brq.2016.02.002.

CHAPTER 13

Jesús Herrera Madueño, Manuel Larrán Jorge, Isabel Martínez Conesa, Domingo Martínez-Martínez, Relationship between corporate social responsibility and competitive performance in Spanish SMEs: Empirical evidence from a stakeholders' perspective, BRQ Business Research Quarterly, Volume 19, Issue 1, 2016, Pages 55-72, ISSN 2340-9436, http://dx.doi.org/10.1016/j.brq.2015.06.002.

INDEX

A

Accounting 2
add functions 42
assumption of heterogeneity 214
Augmented Dickey Fuller (ADF) 152

B

buscadores 20, 21
Business to Business 19
Business to Consumer 19

C

commodities 63
communication mechanisms 307
complementary mediation 334
Control of Corruption (CC) 245
corporate brand identity 177, 178, 179, 180, 183, 184, 185, 187, 188, 193, 194, 195, 196, 197, 198, 202, 352
Corporate Social Responsibility (CSR) 306, 345
customers 30, 31, 33, 38, 39, 48, 49, 50, 51

D

days sales outstanding (DVP) 86
demonstrates 111, 122, 123

E

economically feasible 18
economic order quantity (CEP) 91
economic point 12

e-Learning 42
electronic markets 60
Environmental Kuznets Curvc (EKC) 142, 150

F

financial information 2, 3, 6
Financial Institutions 2
fiscal freedom 237, 239, 240, 249
formal institutional development (FID) 249
fundamental idea 58, 64, 65, 70, 71

G

generated liquidate 68

H

hardware 38, 52, 53
higher risk, higher the return 80
Household Income and Expenditure survey (HIES) 150
Household Income and Expenditure Survey (HIES) 144
hypotheses 12

I

institutional collectivism 254, 260
Internal Rate of Return (IRR) 17
Internet technology 60, 63
investment freedom 237, 241

J

just in time (JIT) 94

L

linear models 1
long-term cost (CPLP) 28
Low Carbon Economy (LCE) 164

M

Material Requirements Planning (PNM) 94
microfoundations 213, 215, 216
Middle East and North African Countries (MENA) 149
Millennium Development Goals (MDGs) 139
multinational enterprises (MNEs) 215

N

National Classification of Economic Activities (CNAE) 282
National Institute of Statistics (INE) 282
Net Present Value (NPV) 17
Network markets 54
neuroticism 112, 117, 118, 122, 123, 127, 128

O

Obviously 24
Other cost factors 95

P

partial least squares (PLS) 306, 338
pendulum 226, 266
performance orientation 254, 257
personology 113, 129
physical products 28, 29, 39, 42, 52
plant and equipment (PPE) 282
political constraints index 235

Political stability and absence of violence/terrorism (PS) 245
Preparation costs 94
proveedora- integration 61
Public Administration 2

R

Regulatory quality (RQ) 245
Rule of law (RL) 245

S

Sketchingdas 73
small and medium enterprises (SMEs) 305, 306
small and medium-size enterprises (SMEs) 217
Social Responsibility (SR) 306
stakeholder management 311, 312, 339
Strategic management 203, 204, 207, 218, 220, 222
Strategic Management 225, 226, 227, 229, 252, 260, 261, 263, 264, 265, 268, 269, 270, 271
Strategic Management arena 227
Strategic Management Journal (SMJ) 205
subsystem 2
supply and demand 30, 54
swings of a pendulum 226, 266

T

traditional economies 53

U

United Nations Development Programme (UNDP) 166

V

Voice and accountability (VA) 245

W

World Commission on Environment and Development (WCED) 141